SpringerBriefs in Applied Sciences and Technology

More information about this series at http://www.springer.com/series/8884

Agnieszka B. Malinowska
Tatiana Odzijewicz · Delfim F.M. Torres

Advanced Methods in the Fractional Calculus of Variations

 Springer

Agnieszka B. Malinowska
Faculty of Computer Science
Białystok University of Technology
Białystok
Poland

Delfim F.M. Torres
Department of Mathematics
University of Aveiro
Aveiro
Portugal

Tatiana Odzijewicz
Department of Mathematics and
  Mathematical Economics
Warsaw School of Economics
Warsaw
Poland

ISSN 2191-530X   ISSN 2191-5318   (electronic)
SpringerBriefs in Applied Sciences and Technology
ISBN 978-3-319-14755-0   ISBN 978-3-319-14756-7   (eBook)
DOI 10.1007/978-3-319-14756-7

Library of Congress Control Number: 2014960038

Springer Cham Heidelberg New York Dordrecht London

Printed on acid-free paper

Springer International Publishing AG Switzerland is part of Springer Science+Business Media
(www.springer.com)

Preface

Fractional differentiation means "differentiation of arbitrary order." Its origin goes back to more than 300 years, when in 1695 L'Hopital asked Leibniz the meaning of $\frac{d^n y}{dx^n}$ for $n = \frac{1}{2}$. After that, many famous mathematicians, like J. Fourier, N.H. Abel, J. Liouville, B. Riemann, among others, contributed to the development of Fractional Calculus (Hilfer 2000; Podlubny 1999; Samko et al. 1993). It is possible to introduce fractional differentiation in several different ways, e.g., by following Riemann–Liouville, Grünwald–Letnikov, Caputo or Miller–Ross. During three centuries, the theory of fractional derivatives has developed as a pure theoretical field of mathematics. In the last few decades, however, it has been shown that fractional differentiation can be useful in various fields: physics (classic and quantum mechanics, thermodynamics, etc.), chemistry, biology, economics, engineering, signal and image processing, and control theory (Baleanu et al. 2012; Machado et al. 2011; Ortigueira 2011).

The calculus of variations and the fractional calculus are connected since the nineteenth century. Indeed, in 1823 Niels Heinrik Abel applied the fractional calculus to the solution of an integral equation that arises in the formulation of the tautochrone problem (Jahanshahi et al. 2015). This problem, sometimes also called the isochrone problem, consists to find the shape of a frictionless wire lying in a vertical plane such that the time of a bead placed on the wire slides to the lowest point of the wire in the same time regardless of where the bead is placed. It turns out that the cycloid is the isochrone as well as the brachistochrone curve, solving simultaneously the brachistochrone problem of the calculus of variations and Abel's fractional problem (Abel 1965). It was however, only in the twentieth century that both areas joined in a unique research field: the fractional calculus of variations.

The fractional calculus of variations consists in extremizing (minimizing or maximizing) functionals whose Lagrangians contain fractional integrals and/or derivatives. It was born in 1996–1997, when Riewe derived Euler–Lagrange fractional differential equations and showed how nonconservative systems in mechanics can be described using fractional derivatives (Riewe 1996, 1997). It is a remarkable result since frictional and nonconservative forces are beyond the usual macroscopic variational treatment and, consequently, beyond the most advanced methods of classical mechanics (Lánczos 1970). Recently, several different approaches have been developed to generalize the least action principle and the Euler–Lagrange equations to include fractional derivatives. Results include problems depending on Caputo fractional derivatives, Riemann–Liouville fractional derivatives, Riesz fractional derivatives and others (Agrawal 2002; Almeida 2012; Almeida et al. 2010, 2012a, b; Almeida and Torres 2009; Bastos et al. 2011a, b; Blaszczyk et al. 2011; Bourdin 2013; Cresson 2007; Frederico and Torres 2008, 2010; Klimek 2005; Lazo and Torres 2013; Malinowska 2012; Malinowska and Torres 2010; Mozyrska and Torres 2010, 2011; Odzijewicz et al. 2012b; Odzijewicz and Torres 2011; Sha et al. 2010). For the state of the art of the fractional calculus of variations, we refer the reader to (Malinowska and Torres 2012); for the numerical aspects see the recent book (Almeida et al. 2015).

A more general unifying perspective to the subject is, however, possible, by considering fractional operators depending on general kernels (Agrawal 2010; Klimek and Lupa 2013; Odzijewicz et al. 2012a, c). In this work we follow such an approach, developing a generalized fractional calculus of variations (Odzijewicz and Torres 2014). We consider variational problems containing generalized fractional integrals and derivatives and study them using standard (indirect) and direct methods. In particular, we obtain necessary optimality conditions of Euler–Lagrange type for the fundamental and isoperimetric problems, natural boundary conditions, and Noether theorems. Existence of solutions is shown under Tonelli type conditions. Moreover, we apply our results to prove existence of eigenvalues, and corresponding orthogonal eigenfunctions, to fractional Sturm–Liouville problems.

Keywords Calculus of variations · Necessary optimality conditions of Euler–Lagrange type · Direct methods · Isoperimetric problem · Noether's theorem · Fractional calculus · Sturm–Liouville problem

Białystok, January 2015   Agnieszka B. Malinowska
Warsaw   Tatiana Odzijewicz
Aveiro   Delfim F.M. Torres

References

Abel NH (1965) Euvres completes de Niels Henrik Abel. Christiana: Imprimerie de Grondahl and Son, New York and London
Agrawal OP (2002) Formulation of Euler–Lagrange equations for fractional variational problem. J Math Anal Appl 272(1):368–379
Agrawal OP (2010) Generalized variational problems and Euler–Lagrange equations. Comput Math Appl 59(5):1852–1864
Almeida R (2012) Fractional variational problems with the Riesz-Caputo derivative. Appl Math Lett 25(2):142–148

Almeida R, Torres DFM (2009) Calculus of variations with fractional derivatives and fractional integrals. Appl Math Lett 22(12):1816–1820

Almeida R, Malinowska AB, Torres DFM (2010) A fractional calculus of variations for multiple integrals with application to vibrating string. J Math Phys 51(3):033503, 12 pp

Almeida R, Malinowska AB, Torres DFM (2012a) Fractional Euler–Lagrange differential equations via Caputo derivatives. In: Fractional Dynamics and Control. Springer, New York, Part 2, pp 109–118

Almeida R, Pooseh S, Torres DFM (2012b) Fractional variational problems depending on indefinite integrals. Nonlinear Anal 75(3):1009–1025

Almeida R, Pooseh S, Torres DFM (2015) Computational methods in the fractional calculus of variations. Imperial College Press, London

Baleanu D, Diethelm K, Scalas E, Trujillo JJ (2012) Fractional calculus. World Scientific Publishing Co. Pte. Ltd., Hackensack

Bastos NRO, Ferreira RAC, Torres DFM (2011a) Necessary optimality conditions for fractional difference problems of the calculus of variations. Discrete Contin Dyn Syst 29(2):417–437

Bastos NRO, Ferreira RAC, Torres DFM (2011b) Discrete-time fractional variational problems. Signal Process 91(3):513–524

Blaszczyk T, Ciesielski M, Klimek M, Leszczynski J (2011) Numerical solution of fractional oscillator equation. Appl Math and Comput 218(6):2480–2488

Bourdin L (2013) Existence of a weak solution for fractional Euler–Lagrange equations. J Math Anal Appl 399(1):239–251

Cresson J (2007) Fractional embedding of differential operators and Lagrangian systems. J Math Phys 48(3):033504, 34 pp

Frederico GSF, Torres DFM (2008) Fractional conservation laws in optimal control theory. Nonlinear Dyn 53(3):215–222

Frederico GSF, Torres DFM (2010) Fractional Noether's theorem in the Riesz-Caputo sense. Appl Math Comput 217(3):1023–1033

Hilfer R (2000) Applications of fractional calculus in physics. World Scientific Publishing, River Edge

Jahanshahi S, Babolian E, Torres DFM, Vahidi A (2015) Solving Abel integral equations of first kind via fractional calculus. J King Saud Univ Sci (in press)

Klimek M (2005) Lagrangian fractional mechanics—a noncommutative approach. Czech J Phys 55(11):1447–1453

Klimek M, Lupa M (2013) Reflection symmetric formulation of generalized fractional variational calculus. Fract Calc Appl Anal 16(1):243–261

Lánczos C (1970) The variational principles of mechanics. vol 4, 4th edn. Mathematical Expositions. University of Toronto Press, Toronto

Lazo MJ, Torres DFM (2013) The DuBois-Reymond fundamental lemma of the fractional calculus of variations and an Euler–Lagrange equation involving only derivatives of Caputo. J Optim Theory Appl 156(1):56–67

Machado JT, Kiryakova V, Mainardi F (2011) Recent history of fractional calculus. Commun Nonlinear Sci Numer Simul 16(3):1140–1153

Malinowska AB (2012) A formulation of the fractional Noether-type theorem for multidimensional Lagrangians. Appl Math Lett 25(11):1941–1946

Malinowska AB, Torres DFM (2010) Generalized natural boundary conditions for fractional variational problems in terms of the Caputo derivative. Comput Math Appl 59(9):3110–3116

Malinowska AB, Torres DFM (2012) Introduction to the fractional calculus of variations. Imperial College Press, London

Mozyrska D, Torres DFM (2010) Minimal modified energy control for fractional linear control systems with the Caputo derivative. Carpathian J Math 26(2):210–221

Mozyrska D, Torres DFM (2011) Modified optimal energy and initial memory of fractional continuous-time linear systems. Signal Process 91(3):379–385

Odzijewicz T, Malinowska AB, Torres DFM (2012a) Generalized fractional calculus with applications to the calculus of variations. Comput Math Appl 64(10):3351–3366

Odzijewicz T, Malinowska AB, Torres DFM (2012b) Fractional variational calculus with classical and combined Caputo derivatives. Nonlinear Anal 75(3):1507–1515

Odzijewicz T, Malinowska AB, Torres DFM (2012c) Fractional calculus of variations in terms of a generalized fractional integral with applications to physics. Abstr Appl Anal, 2012(871912), 24 pp

Odzijewicz T, Torres DFM (2011) Fractional calculus of variations for double integrals. Balkan J Geom Appl 16(2):102–113

Odzijewicz T, Torres DFM (2014) The generalized fractional calculus of variations. Southeast Asian Bull Math 38(1):93–117

Ortigueira MD (2011) Fractional calculus for scientists and engineers. Lecture Notes in Electrical Engineering, vol 84. Springer, Dordrecht

Podlubny I (1999) Fractional differential equations. Academic Press, San Diego

Riewe F (1996) Nonconservative Lagrangian and Hamiltonian mechanics. Phys Rev E (3) 53 (2):1890–1899

Riewe F (1997) Mechanics with fractional derivatives. Phys Rev E (3), part B 55(3):3581–3592

Samko SG, Kilbas AA, Marichev OI (1993) Fractional integrals and derivatives. Gordon and Breach, Yverdon

Sha Z, Jing-Li F, Yong-Song L (2010) Lagrange equations of nonholonomic systems with fractional derivatives. Chin Phys B 19(12):120301, 5 pp

Acknowledgments

This work was supported by Portuguese funds through the *Center for Research and Development in Mathematics and Applications* (CIDMA), and *The Portuguese Foundation for Science and Technology* (FCT), within project UID/MAT/04106/ 2013. Malinowska was also supported by Bialystok University of Technology; Odzijewicz by FCT post-doc fellowship SFRH/BPD/92691/2013; Torres by project OCHERA, PTDC/EEI-AUT/1450/2012, co-financed by FEDER under POFC-QREN with COMPETE reference FCOMP-01-0124-FEDER-028894, and EU funding under the 7th Framework Programme FP7-PEOPLE-2010-ITN, grant agreement number 264735-SADCO.

Any comments or suggestions related to the material here contained are more than welcome, and may be submitted by post or by electronic mail to the authors:

Agnieszka B. Malinowska <a.malinowska@pb.edu.pl>
Department of Mathematics, Faculty of Computer Science
Bialystok University of Technology
15-351 Białystok, Poland

Tatiana Odzijewicz <tatiana.odzijewicz@gmail.com>
Department of Mathematics and Mathematical Economics
Warsaw School of Economics
al. Niepodleglosci 162, 02-554, Warsaw, Poland

Delfim F.M. Torres <delfim@ua.pt>
Center for Research and Development in Mathematics and Applications
Department of Mathematics, University of Aveiro
3810-193 Aveiro, Portugal

Contents

1	Introduction	1
	References	5

2	Fractional Calculus	7
	2.1 One-Dimensional Fractional Calculus	8
	2.1.1 Classical Fractional Operators	8
	2.1.2 Variable Order Fractional Operators	12
	2.1.3 Generalized Fractional Operators	14
	2.2 Multidimensional Fractional Calculus	16
	2.2.1 Classical Partial Fractional Integrals and Derivatives	16
	2.2.2 Variable Order Partial Fractional Integrals and Derivatives	17
	2.2.3 Generalized Partial Fractional Operators	19
	References	20

3	Fractional Calculus of Variations	23
	3.1 Fractional Euler–Lagrange Equations	24
	3.2 Fractional Embedding of Euler–Lagrange Equations	27
	References	29

4	Standard Methods in Fractional Variational Calculus	31
	4.1 Properties of Generalized Fractional Integrals	33
	4.1.1 Boundedness of Generalized Fractional Operators	33
	4.1.2 Generalized Fractional Integration by Parts	35
	4.2 Fundamental Problem	37
	4.3 Free Initial Boundary	45
	4.4 Isoperimetric Problem	47
	4.5 Noether's Theorem	56
	4.6 Variational Calculus in Terms of a Generalized Integral	60
	4.7 Generalized Variational Calculus of Several Variables	63

4.7.1   Multidimensional Generalized Fractional Integration
        by Parts   64
4.7.2   Fundamental Problem.   67
4.7.3   Dirichlet's Principle   72
4.7.4   Isoperimetric Problem   73
4.7.5   Noether's Theorem   77
4.8   Conclusion   80
References   81

5   **Direct Methods in Fractional Calculus of Variations**   83
5.1   Existence of a Minimizer for a Generalized Functional   83
5.1.1   A Tonelli-Type Theorem   84
5.1.2   Sufficient Condition for Regular Lagrangians   86
5.1.3   Sufficient Condition for Coercive Functionals   87
5.1.4   Examples of Lagrangians   88
5.2   Necessary Optimality Condition for a Minimizer   90
5.3   Some Improvements.   93
5.3.1   A First Weaker Convexity Assumption   93
5.3.2   A Second Weaker Convexity Assumption.   95
5.4   Conclusion   96
References   96

6   **Application to the Sturm–Liouville Problem**   99
6.1   Useful Lemmas   100
6.2   The Fractional Sturm–Liouville Problem.   104
6.2.1   Existence of Discrete Spectrum   105
6.2.2   The First Eigenvalue   116
6.2.3   An Illustrative Example   119
References   121

7   **Conclusion**   123
References   124

Appendix A: Two Convergence Lemmas   127

Index   133

Chapter 1
Introduction

Abstract This book is dedicated to the generalized fractional calculus of variations and its main task is to unify and extend results concerning the standard fractional variational calculus.

Keywords Fractional calculus · Calculus of variations · Fractional calculus of variations · Euler–Lagrange equation · Nonconservative forces

The calculus of variations is a mathematical research field that was born in 1696 with the solution to the brachistochrone problem (see, e.g., van Brunt 2004) and is focused on finding extremal values of functionals—see, e.g., (Giaquinta and Hildebrandt 2004; van Brunt 2004). Usually, considered functionals are given in the form of an integral that involves an unknown function and its derivatives. Variational problems are particularly attractive because of their manyfold applications, e.g., in physics, engineering, and economics. The variational integral may represent an action, energy, or cost functional (Weinstock 1952). The calculus of variations possesses also important connections with other fields of mathematics, e.g., with fractional calculus.

Fractional calculus, i.e., the calculus of noninteger order derivatives, has also its origin in the 1600s. It is a generalization of (integer) differential calculus, allowing to define derivatives (and integrals) of real or complex order (Kilbas et al. 2006; Podlubny 1999; Samko et al. 1993). In the last few decades, fractional problems have received increasing attention of many researchers. As mentioned in (Balachandran et al. 2012), *Science Watch of Thomson Reuters* identified the subject as an *Emerging Research Front* area. Fractional derivatives are nonlocal operators and are historically applied in the study of nonlocal or time-dependent processes (Podlubny 1999). The first and well-established application of fractional calculus in physics was in the framework of anomalous diffusion, which is related to features observed in many physical systems. Here, we can mention the report (Metzler and Klafter 2000) demonstrating that fractional equations work as a complementary tool in the description of anomalous transport processes. Within the fractional approach it is possible to include external fields in a straightforward manner. As a consequence, in a short period of time the list of applications was expanded. Applications include chaotic dynamics (Zaslavsky 2008), material sciences (Mainardi 2010),

© The Author(s) 2015

A.B. Malinowska et al., *Advanced Methods in the Fractional Calculus
of Variations*, SpringerBriefs in Applied Sciences and Technology,
DOI 10.1007/978-3-319-14756-7_1

mechanics of fractal and complex media (Carpinteri and Mainardi 1997; Li and
Ostoja-Starzewski 2011), quantum mechanics (Hilfer 2000; Laskin 2000), physical
kinetics (Zaslavsky and Edelman 2004), long-range dissipation (Tarasov 2006b), and
long-range interaction (Tarasov 2006a; Tarasov and Zaslavsky 2006), just to mention a few. This diversity of applications makes the fractional calculus an important
subject, which requires serious attention and strong interest. The fractional calculus
begins with a definition of fractional operator. Let

$$_aI_t^1 x(t) := \int_a^t x(\tau)\, d\tau.$$

It is easy to prove, by induction, that

$$_aI_t^n x(t) = \frac{1}{(n-1)!} \int_a^t (t-\tau)^{n-1} x(\tau)\, d\tau$$

for all $n \in \mathbb{N}$: if it is true for the n-fold integral, then

$$_aI_t^{n+1} x(t) = {}_aI_t^1 \left(\frac{1}{(n-1)!} \int_a^t (t-\tau)^{n-1} x(\tau)\, d\tau \right)$$

$$= \int_a^t \left(\frac{1}{(n-1)!} \int_a^\xi (\xi-\tau)^{n-1} x(\tau)\, d\tau \right) d\xi.$$

Interchanging the order of integration gives

$$_aI_t^{n+1} x(t) = \frac{1}{n!} \int_a^t (t-\tau)^n x(\tau)\, d\tau.$$

The (left Riemann–Liouville) fractional integral of $x(t)$ of order $\alpha > 0$, is then
defined with the help of Euler's Gamma function Γ:

$$_aI_t^\alpha x(t) := \frac{1}{\Gamma(\alpha)} \int_a^t (t-\tau)^{\alpha-1} x(\tau)\, d\tau. \tag{1.1}$$

If m is the smallest integer exceeding α, then we define the *fractional Riemann–
Liouville derivative of f* of order α as

$$_aD_x^\alpha f(x) = \frac{d^m}{dx^m}\left[_aI_x^{m-\alpha}f(x)\right] = \frac{1}{\Gamma(m-\alpha)}\frac{d^m}{dx^m}\int_a^x f(t)(x-t)^{m-\alpha-1}\,dt. \quad (1.2)$$

Another definition of fractional derivative was introduced by M. Caputo in 1967, interchanging the order of the operators $\frac{d^m}{dx^m}$ and $_aI_x^{m-\alpha}$ in (1.2):

$$_a^CD_x^\alpha := {_aI_x^{m-\alpha}} \circ \frac{d^m}{dx^m}. \quad (1.3)$$

In this book we shall consider generalizations of operators (1.1), (1.2) and (1.3) by considering more general kernels.

The classical fundamental problem of the calculus of variations is formulated as follows: minimize (or maximize) the functional

$$\mathcal{J}(x) = \int_a^b L(t, x(t), x'(t))\,dt$$

on $\mathcal{D} = \{x \in C^1([a, b]) : x(a) = x_a, x(b) = x_b\}$, where $L : [a, b] \times \mathbb{R}^{2n} \to \mathbb{R}$ is twice continuously differentiable. In mechanics, function L is called the *Lagrangian*; functional $\mathcal{J}$ is called the *action*. If x gives a (local) minimum (or maximum) to $\mathcal{J}$ on $\mathcal{D}$, then $\frac{d}{dt}\partial_3 L\left(t, x(t), x'(t)\right) = \partial_2 L\left(t, x(t), x'(t)\right)$ holds for all $t \in [a, b]$, where we are using the notation $\partial_i F$ for the partial derivative of a function F with respect to its ith argument. This is the celebrated Euler–Lagrange equation, which is a first-order necessary optimality condition. In mechanics, if Lagrangian L does not depend explicitly on t, then the *energy* $\mathcal{E}(x) := -L(x, x') + \frac{\partial L}{\partial x'}(x, x') \cdot x'$ is constant along physical trajectories x (that is, along the solutions of the Euler–Lagrange equations). In real systems, friction corrupts conservation of energy, and the usefulness of variational principles is lost: "forces of a frictional nature are outside the realm of variational principles." For conservative systems, variational methods are equivalent to the original method used by Newton. However, while Newton's equations allow nonconservative forces, the later techniques of Lagrangian and Hamiltonian mechanics have no direct way to deal with them: *frictional and nonconservative forces are beyond the usual macroscopic variational treatment and, consequently, beyond the most advanced methods of classical mechanics* (Lánczos 1970). In 1931, Bauer proved that it is impossible to use a variational principle in order to derive a single linear dissipative equation of motion with constant coefficients. Bauer's theorem expresses the well-known belief that there is no direct method of applying variational principles to nonconservative systems, which are characterized by friction or other dissipative processes. Fractional derivatives provide an elegant solution to the problem. Indeed, the proof of Bauer's theorem relies on the tacit assumption that all derivatives are of integer order. If a Lagrangian is constructed using fractional (noninteger order) derivatives, then the resulting equation of motion can be nonconservative! This was first proved by F. Riewe in 1996/1997 (Riewe 1996, 1997), marking the beginning of the *Fractional Calculus of Variations* (FCV). Because most processes observed in the physical world are nonconservative, FCV constitutes an important research area, allowing to apply the power of variational methods to real systems. The first book on the subject is (Malinowska and Torres 2012), which provides a gentle introduction to the FCV. The numerical aspects are treated in Almeida et al. (2015). The model problem considered in Malinowska and Torres (2012) is to find an admissible function giving a minimum value to an integral functional that depends on an unknown function (or functions), of one or several variables, and its fractional derivatives and/or fractional integrals. Here we explain how the main results presented in Malinowska and Torres (2012) can be extended by considering generalized fractional operators (Odzijewicz 2013). We consider problems where the Lagrangians depend not only on classical derivatives but also on generalized fractional operators. Moreover, we discuss even more general problems, where also classical integrals are substituted by generalized fractional integrals and obtain general theorems, for several types of variational problems, which are valid for rather arbitrary operators and kernels. As special cases, one obtains the recent results available in the literature of fractional variational calculus (Herrera et al. 1986; Klimek 2009; Malinowska and Torres 2012).

The book is organized as follows. We begin by giving preliminary definitions and properties of fractional operators under consideration (Chap. 2). Moreover, we briefly describe recent results on the fractional calculus of variations (Chap. 3). In Chap. 4 we apply standard methods to solve several problems of the generalized fractional calculus of variations. We consider problems with Lagrangians depending on classical derivatives, generalized fractional integrals, and generalized fractional derivatives. We obtain necessary optimality conditions for the basic and isoperimetric problems, as well as natural boundary conditions for free boundary value problems. In addition, we prove a generalized fractional counterpart of Noether's theorem without transformation of time. We consider the case of one and several independent variables. Moreover, each section contains illustrative optimization problems. Chapter 5 is dedicated to direct methods in the fractional calculus of variations. We prove a generalized fractional Tonelli's theorem, showing existence of minimizers for fractional variational functionals. Then we obtain necessary optimality conditions for minimizers. Several illustrative examples are presented. In the last Chap. 6 we show a certain application of the fractional variational calculus. More precisely, we prove existence of eigenvalues and corresponding eigenfunctions for the fractional Sturm–Liouville problem using variational methods. Moreover, we show two theorems concerning the lowest eigenvalue and illustrate our results through an example. We end the book with a conclusion, pointing out some directions for future research.

This small book is the fruit of several years of research by the authors, and the results first appeared in peer-reviewed international journals, as chapters in books, or in conference proceedings (Bourdin et al. 2013, 2014; Klimek et al. 2014; Odzijewicz et al. 2012a, b, c, d, 2013a, b, c, d, e; Odzijewicz and Torres 2012, 2014).

References

Almeida R, Pooseh S, Torres DFM (2015) Computational methods in the fractional calculus of variations. Imperial College Press, London

Balachandran K, Park JY, Trujillo JJ (2012) Controllability of nonlinear fractional dynamical systems. Nonlinear Anal 75(4):1919–1926

Bourdin L, Odzijewicz T, Torres DFM (2013) Existence of minimizers for fractional variational problems containing Caputo derivatives. Adv Dyn Syst Appl 8(1):3–12

Bourdin L, Odzijewicz T, Torres DFM (2014) Existence of minimizers for generalized Lagrangian functionals and a necessary optimality condition—application to fractional variational problems. Differ Integral Equ 27(7–8):743–766

Carpinteri A, Mainardi F (1997) Fractals and fractional calculus in continuum mechanics, vol 378. CISM courses and lectures. Springer, Vienna

Giaquinta M, Hildebrandt S (2004) Calculus of variations I. Springer, Berlin

Herrera L, Nunez L, Patino A, Rago H (1986) A variational principle and the classical and quantum mechanics of the damped harmonic oscillator. Am J Phys 54(3):273–277

Hilfer R (2000) Applications of fractional calculus in physics. World Scientific Publishing, River Edge

Kilbas AA, Srivastava HM, Trujillo JJ (2006) Theory and applications of fractional differential equations, vol 204. North-Holland mathematics studies. Elsevier, Amsterdam

Klimek M (2009) On solutions of linear fractional differential equations of a variational type. The Publishing Office of Czestochowa University of Technology, Czestochowa

Klimek M, Odzijewicz T, Malinowska AB (2014) Variational methods for the fractional Sturm-Liouville problem. J Math Anal Appl 416(1):402–426

Lánczos C (1970) The variational principles of mechanics, vol 4. 4th edn. Mathematical expositions. University of Toronto Press, Toronto

Laskin N (2000) Fractional quantum mechanics and Lévy path integrals. Phys Lett A 268(4–6):298–305

Li J, Ostoja-Starzewski M (2011) Micropolar continuum mechanics of fractal media. Intern J Eng Sci 49(12):1302–1310

Mainardi F (2010) Fractional calculus and waves in linear viscoelasticity. Imperial College Press, London

Malinowska AB, Torres DFM (2012) Introduction to the fractional calculus of variations. Imperial College Press, London

Metzler R, Klafter J (2000) The random walk's guide to anomalous diffusion: a fractional dynamics approach. Phys Rep 339(1):77

Odzijewicz T (2013) Generalized fractional calculus of variations. PhD Thesis, University of Aveiro

Odzijewicz T, Malinowska AB, Torres DFM (2012a) Generalized fractional calculus with applications to the calculus of variations. Comput Math Appl 64(10):3351–3366

Odzijewicz T, Malinowska AB, Torres DFM (2012b) Fractional variational calculus with classical and combined Caputo derivatives. Nonlinear Anal 75(3):1507–1515

Odzijewicz T, Malinowska AB, Torres DFM (2012c) Fractional calculus of variations in terms of a generalized fractional integral with applications to physics. Abstr Appl Anal 2012(871912):24

Odzijewicz T, Malinowska AB, Torres DFM (2012d) Variable order fractional variational calculus for double integrals. In: Proceedings of the IEEE conference on decision and control, vol 6426489, pp 6873–6878

Odzijewicz T, Malinowska AB, Torres DFM (2013a) Fractional variational calculus of variable order. Advances in harmonic analysis and operator theory, Operator theory: advances and applications, vol 229. Birkhäuser, Basel, pp 291–301

Odzijewicz T, Malinowska AB, Torres DFM (2013b) Green's theorem for generalized fractional derivative. Fract Calc Appl Anal 16(1):64–75

Odzijewicz T, Malinowska AB, Torres DFM (2013c) A generalized fractional calculus of variations. Control Cybern 42(2):443–458

Odzijewicz T, Malinowska AB, Torres DFM (2013d) Fractional calculus of variations of several independent variables. Eur Phys J Spec Top 222(8):1813–1826

Odzijewicz T, Malinowska AB, Torres DFM (2013e) Noether's theorem for fractional variational problems of variable order. Cent Eur J Phys 11(6):691–701

Odzijewicz T, Torres DFM (2012) Calculus of variations with classical and fractional derivatives. Math Balkanica 26(1–2):191–202

Odzijewicz T, Torres DFM (2014) The generalized fractional calculus of variations. Southeast Asian Bull Math 38(1):93–117

Podlubny I (1999) Fractional differential equations. Academic Press, San Diego

Riewe F (1996) Nonconservative Lagrangian and Hamiltonian mechanics. Phys Rev E (3) 53(2):1890–1899

Riewe F (1997) Mechanics with fractional derivatives. Phys Rev E (3), part B 55(3):3581–3592

Samko SG, Kilbas AA, Marichev OI (1993) Fractional integrals and derivatives. Gordon and Breach, Yverdon

Tarasov VE (2006a) Continuous limit of discrete systems with long-range interaction. J Phys A 39(48):14895–14910

Tarasov VE (2006b) Fractional statistical mechanics. Chaos 16(3), 033108, 7 pp

Tarasov VE, Zaslavsky GM (2006) Fractional dynamics of coupled oscillators with long-range interaction. Chaos 16(2), 023110, 13 pp

van Brunt B (2004) The calculus of cariations. Springer, New York

Weinstock R (1952) Calculus of variations with applications to physics and engineering. McGraw-Hill Book Company Inc, New York

Zaslavsky GM (2008) Hamiltonian chaos and fractional dynamics. Reprint of the 2005 original. Oxford University Press, Oxford

Zaslavsky GM, Edelman MA (2004) Fractional kinetics: from pseudochaotic dynamics to Maxwell's demon. Phys D 193(1–4):128–147

Chapter 2
Fractional Calculus

Abstract A brief exposition of fractional order operators and their properties is given. After that, we introduce the notion of generalized fractional operators.

Keywords Fractional derivatives and integrals · Generalized fractional derivatives and integrals · Fractional derivatives and integrals of variable order · Riemann-Liouville, Hadamard and Caputo operators · Fractional integration by parts · Multi-dimensional generalized fractional calculus

Fractional calculus was introduced on September 30, 1695. On that day, Leibniz wrote a letter to L'Hôpital, raising the possibility of generalizing the meaning of derivatives from integer order to noninteger order derivatives. L'Hôpital wanted to know the result for the derivative of order $n = 1/2$. Leibniz replied that *"one day, useful consequences will be drawn"* and, in fact, his vision became a reality. However, the study of noninteger order derivatives did not appear in the literature until 1819, when Lacroix presented a definition of fractional derivative based on the usual expression for the nth derivative of the power function (Lacroix 1819). Within years the fractional calculus became a very attractive subject to mathematicians, and many different forms of fractional (i.e., noninteger) differential operators were introduced: the Grunwald–Letnikow, Riemann–Liouville, Hadamard, Caputo, Riesz (Hilfer 2000; Kilbas et al. 2006; Podlubny 1999; Samko et al. 1993) and the more recent notions of Cresson (2007), Katugampola (2011), Klimek (2005), Kilbas and Saigo (2004) or variable order fractional operators introduced by Samko and Ross (1993).

In 2010, an interesting perspective to the subject, unifying all mentioned notions of fractional derivatives and integrals, was introduced in Agrawal (2010) and later studied in Bourdin et al. (2014), Klimek and Lupa (2013), Odzijewicz et al. (2012a, b, 2013a, b, c). Precisely, authors considered general operators, which by choosing special kernels, reduce to the standard fractional operators. However, other nonstandard kernels can also be considered as particular cases.

This chapter presents preliminary definitions and facts of classical, variable order, and generalized fractional operators.

A.B. Malinowska et al., *Advanced Methods in the Fractional Calculus
of Variations*, SpringerBriefs in Applied Sciences and Technology,
DOI 10.1007/978-3-319-14756-7_2

2.1 One-Dimensional Fractional Calculus

We begin with basic facts on the one-dimensional classical, variable order, and generalized fractional operators.

2.1.1 Classical Fractional Operators

In this section, we present definitions and properties of the one-dimensional fractional integrals and derivatives under consideration. The reader interested in the subject is refereed to the books (Kilbas et al. 2006; Klimek 2009; Podlubny 1999; Samko et al. 1993).

Definition 2.1 (*Left and right Riemann–Liouville fractional integrals*) We define the left and the right Riemann–Liouville fractional integrals $_aI_t^\alpha$ and $_tI_b^\alpha$ of order $\alpha \in \mathbb{R}$ $(\alpha > 0)$ by

$$_aI_t^\alpha[f](t) := \frac{1}{\Gamma(\alpha)} \int_a^t \frac{f(\tau)\,\mathrm{d}\tau}{(t-\tau)^{1-\alpha}}, \quad t \in (a,b], \tag{2.1}$$

and

$$_tI_b^\alpha[f](t) := \frac{1}{\Gamma(\alpha)} \int_t^b \frac{f(\tau)\,\mathrm{d}\tau}{(\tau-t)^{1-\alpha}}, \quad t \in [a,b), \tag{2.2}$$

respectively. Here $\Gamma(\alpha)$ denotes Euler's Gamma function. Note that, $_aI_t^\alpha[f]$ and $_tI_b^\alpha[f]$ are defined a.e. on (a,b) for $f \in L^1(a,b;\mathbb{R})$.

One can also define fractional integral operators in the frame of Hadamard setting. In the following, we present definitions of Hadamard fractional integrals.

Definition 2.2 (*Left and right Hadamard fractional integrals*) Let $0 \le a < b < \infty$. We define the left-sided and right-sided Hadamard integrals of fractional order $\alpha \in \mathbb{R}$ $(\alpha > 0)$ by

$$_aJ_t^\alpha[f](t) := \frac{1}{\Gamma(\alpha)} \int_a^t \left(\log \frac{t}{\tau}\right)^{\alpha-1} \frac{f(\tau)\,\mathrm{d}\tau}{\tau}, \quad t > a$$

and

$$_tJ_b^\alpha[f](t) := \frac{1}{\Gamma(\alpha)} \int_t^b \left(\log \frac{\tau}{t}\right)^{\alpha-1} \frac{f(\tau)\,\mathrm{d}\tau}{\tau}, \quad t < b,$$

respectively.

Since it is enough for the purposes of this book, we define Riemann–Liouville fractional derivatives of order α with $0 < \alpha < 1$. A more general definition for any α with $Re(\alpha) > 0$ can be found in Kilbas et al. (2006).

Definition 2.3 (*Left and right Riemann–Liouville fractional derivatives*) The left Riemann–Liouville fractional derivative of order $\alpha \in \mathbb{R}$ ($0 < \alpha < 1$) of a function f, denoted by $_aD_t^\alpha[f]$, is defined by

$$\forall t \in (a, b), \quad _aD_t^\alpha[f](t) := \frac{d}{dt}\,_aI_t^{1-\alpha}[f](t).$$

Similarly, the right Riemann–Liouville fractional derivative of order α of a function f, denoted by $_tD_b^\alpha[f]$, is defined by

$$\forall t \in [a, b), \quad _tD_b^\alpha[f](t) := -\frac{d}{dt}\,_tI_b^{1-\alpha}[f](t).$$

As we can see below, Riemann–Liouville fractional integral and differential operators of power functions return power functions.

Property 2.4 (Property 2.1 (Kilbas et al. 2006)) *Let $\alpha, \beta > 0$. Then the following identities hold:*

$$_aI_t^\alpha[(\tau - a)^{\beta-1}](t) = \frac{\Gamma(\beta)}{\Gamma(\beta + \alpha)}(t - a)^{\beta+\alpha-1},$$

$$_aD_t^\alpha[(\tau - a)^{\beta-1}](t) = \frac{\Gamma(\beta)}{\Gamma(\beta - \alpha)}(t - a)^{\beta-\alpha-1},$$

$$_tI_b^\alpha[(b - \tau)^{\beta-1}](t) = \frac{\Gamma(\beta)}{\Gamma(\beta + \alpha)}(b - t)^{\beta+\alpha-1},$$

and

$$_tD_b^\alpha[(b - \tau)^{\beta-1}](t) = \frac{\Gamma(\beta)}{\Gamma(\beta - \alpha)}(b - t)^{\beta-\alpha-1}.$$

Definition 2.5 (*Left and right Caputo fractional derivatives*) The left and the right Caputo fractional derivatives of order $\alpha \in \mathbb{R}$ ($0 < \alpha < 1$) are given by

$$\forall t \in (a, b], \quad _a^C D_t^\alpha[f](t) := _aI_t^{1-\alpha}\left[\frac{d}{dt}f\right](t)$$

and

$$\forall t \in [a, b), \quad _t^C D_b^\alpha[f](t) := -_tI_b^{1-\alpha}\left[\frac{d}{dt}f\right](t),$$

respectively.

Let $0 < \alpha < 1$ and $f \in AC([a, b]; \mathbb{R})$, where AC denotes the class of absolutely continuous functions. Then the Riemann–Liouville and Caputo fractional derivatives satisfy relations

$$\,_a^C D_t^\alpha[f](t) = \,_a D_t^\alpha[f](t) - \frac{f(a)}{(t-a)^\alpha \Gamma(1-\alpha)}, \tag{2.3}$$

$$\,_t^C D_b^\alpha[f](t) = -\,_t D_b^\alpha[f](t) + \frac{f(b)}{(b-t)^\alpha \Gamma(1-\alpha)}, \tag{2.4}$$

that can be found in Kilbas et al. (2006). Moreover, for Riemann–Liouville fractional integrals and derivatives, the following composition rules hold:

$$\left(\,_a I_t^\alpha \circ \,_a D_t^\alpha\right)[f](t) = f(t), \tag{2.5}$$

$$\left(\,_t I_b^\alpha \circ \,_t D_b^\alpha\right)[f](t) = f(t), \tag{2.6}$$

provided that $f \in L^1(a, b; \mathbb{R})$, $\,_a I_t^\alpha[f]$, $\,_t I_b^\alpha[f] \in AC([a, b]; \mathbb{R})$ and $\,_a I_t^\alpha f(a) = 0$, $\,_t I_b^\alpha f(b) = 0$. Note that, if $f(a) = 0$, then (2.3) and (2.5) give

$$\left(\,_a I_t^\alpha \circ \,_a^C D_t^\alpha\right)[f](t) = \left(\,_a I_t^\alpha \circ \,_a D_t^\alpha\right)[f](t) = f(t), \tag{2.7}$$

and if $f(b) = 0$, then (2.4) and (2.6) imply that

$$\left(\,_t I_b^\alpha \circ \,_t^C D_b^\alpha\right)[f](t) = \left(\,_t I_b^\alpha \circ \,_t D_b^\alpha\right)[f](t) = f(t). \tag{2.8}$$

The following assertion shows that Riemann–Liouville fractional integrals satisfy semigroup property.

Property 2.6 (Lemma 2.3 (Kilbas et al. 2006)) *Let α, $\beta > 0$ and $f \in L^r(a, b; \mathbb{R})$ $(1 \leq r \leq \infty)$. Then, equations*

$$\left(\,_a I_t^\alpha \circ \,_a I_t^\beta\right)[f](t) = \,_a I_t^{\alpha+\beta}[f](t)$$

and

$$\left(\,_t I_b^\alpha \circ \,_t I_b^\beta\right)[f](t) = \,_t I_b^{\alpha+\beta}[f](t)$$

are satisfied a.e. in (a, b).

Next results show that, for certain classes of functions, Riemann–Liouville fractional derivatives and Caputo fractional derivatives are left inverse operators of Riemann–Liouville fractional integrals.

Property 2.7 (cf. Lemma 2.4 (Kilbas et al. 2006)) *If* $0 < \alpha < 1$ *and* $f \in L^r(a, b; \mathbb{R})$ $(1 \leq r \leq \infty)$, *then the following is true:*

$$\left({}_aD_t^\alpha \circ {}_aI_t^\alpha\right)[f](t) = f(t),$$

$$\left({}_tD_b^\alpha \circ {}_tI_b^\alpha\right)[f](t) = f(t),$$

a.e. in (a, b).

Property 2.8 (cf. Lemma 2.21 (Kilbas et al. 2006)) *Let* $0 < \alpha < 1$. *If* f *is continuous on the interval* $[a, b]$, *then*

$$\left({}_a^C D_t^\alpha \circ {}_aI_t^\alpha\right)[f](t) = f(t),$$

$$\left({}_t^C D_b^\alpha \circ {}_tI_b^\alpha\right)[f](t) = f(t).$$

For r-Lebesgue integrable functions, Riemann–Liouville fractional integrals and derivatives satisfy the following composition properties:

Property 2.9 (cf. Property 2.2 (Kilbas et al. 2006)) *Let* $0 < \beta < \alpha < 1$ *and* $f \in L^r$ $(a, b; \mathbb{R})$ $(1 \leq r \leq \infty)$. *Then, relations*

$$\left({}_aD_t^\beta \circ {}_aI_t^\alpha\right)[f](t) = {}_aI_t^{\alpha-\beta}[f](t)$$

and

$$\left({}_tD_b^\beta \circ {}_tI_b^\alpha\right)[f](t) = {}_tI_b^{\alpha-\beta}[f](t)$$

are satisfied a.e. in (a, b).

In classical calculus, integration by parts formula relates the integral of a product of functions to the integral of their derivative and antiderivative. As we can see below, this formula works also for fractional derivatives, however, it changes the type of differentiation: left Riemann–Lioville fractional derivatives are transformed to right Caputo fractional derivatives.

Property 2.10 (cf. Lemma 2.19 (Klimek 2009)) *Assume that* $0 < \alpha < 1$, $f \in AC([a, b]; \mathbb{R})$ *and* $g \in L^r(a, b; \mathbb{R})$ $(1 \leq r \leq \infty)$. *Then, the following integration by parts formula holds:*

$$\int_a^b f(t) {}_aD_t^\alpha[g](t)\, dt = \int_a^b g(t) {}_t^C D_b^\alpha[f](t)\, dt + f(t) {}_aI_t^{1-\alpha}[g](t)\Big|_{t=a}^{t=b}. \quad (2.9)$$

Let us recall the following property yielding boundedness of Riemann–Liouville fractional integral in the space $L^r(a, b; \mathbb{R})$ (cf. Lemma 2.1, formula 2.1.23, from the monograph by Kilbas et al. (2006)).

Property 2.11 *The fractional integral $_aI_t^\alpha$ is bounded in the space $L^r(a, b; \mathbb{R})$ for $\alpha \in (0, 1)$ and $r \geq 1$:*

$$\||_aI_t^\alpha[f]\||_{L^r} \leq K_\alpha \||f\||_{L^r}, \quad K_\alpha = \frac{(b-a)^\alpha}{\Gamma(\alpha+1)}. \tag{2.10}$$

2.1.2 Variable Order Fractional Operators

In 1993, Samko and Ross (1993) proposed an interesting generalization of fractional operators. They introduced the study of fractional integration and differentiation when the order is not a constant but a function. Afterwards, several works were dedicated to variable order fractional operators, their applications and interpretations (Almeida and Samko 2009; Coimbra 2003; Lorenzo and Hartley 2002). In particular, Samko's variable order fractional calculus turns out to be very useful in mechanics and in the theory of viscous flows (Coimbra 2003; Diaz and Coimbra 2009; Lorenzo and Hartley 2002; Pedro et al. 2008; Ramirez and Coimbra 2010, 2011). Indeed, many physical processes exhibit fractional order behavior that may vary with time or space (Lorenzo and Hartley 2002). The paper (Coimbra 2003) is devoted to the study of a variable order fractional differential equation that characterizes some problems in the theory of viscoelasticity. In Diaz and Coimbra (2009) the authors analyze the dynamics and control of a nonlinear variable viscoelasticity oscillator, and two controllers are proposed for the variable order differential equations that track an arbitrary reference function. The work (Pedro et al. 2008) investigates the drag force acting on a particle due to the oscillatory flow of a viscous fluid. The drag force is determined using the variable order fractional calculus, where the order of derivative vary according to the dynamics of the flow. In Ramirez and Coimbra (2011) a variable order differential equation for a particle in a quiescent viscous liquid is developed. For more on the application of variable order fractional operators to the modeling of dynamic systems, we refer the reader to the review article (Ramirez and Coimbra 2010).

Let us introduce the following triangle:

$$\Delta := \left\{(t, \tau) \in \mathbb{R}^2 : a \leq \tau < t \leq b\right\},$$

and let $\alpha(t, \tau) : \Delta \to [0, 1]$ be such that $\alpha \in C^1\left(\bar{\Delta}; \mathbb{R}\right)$, where $\bar{\Delta}$ denotes the closure of the set Δ.

Definition 2.12 (*Left and right Riemann–Liouville integrals of variable order*) Operator

$$_aI_t^{\alpha(\cdot,\cdot)}[f](t) := \int\limits_a^t \frac{1}{\Gamma(\alpha(t, \tau))}(t-\tau)^{\alpha(t,\tau)-1} f(\tau)\, d\tau \quad (t > a)$$

is the left Riemann–Liouville integral of variable fractional order $\alpha(\cdot, \cdot)$, while

$$_t I_b^{\alpha(\cdot,\cdot)}[f](t) := \int_t^b \frac{1}{\Gamma(\alpha(\tau,t))} (\tau - t)^{\alpha(\tau,t)-1} f(\tau)\, d\tau \quad (t < b)$$

is the right Riemann–Liouville integral of variable fractional order $\alpha(\cdot, \cdot)$.

The following example gives a variable order fractional integral for the power function $(t - a)^\gamma$.

Example 2.13 (*cf. Equation 4 of* (Samko and Ross 1993)) Let $\alpha(t, \tau) = \alpha(t)$ be a function depending only on variable t, $0 < \alpha(t) < 1$ for almost all $t \in (a, b)$ and $\gamma > -1$. Then,

$$_a I_t^{\alpha(\cdot)} (t - a)^\gamma = \frac{\Gamma(\gamma + 1)(t - a)^{\gamma+\alpha(t)}}{\Gamma(\gamma + \alpha(t) + 1)}. \tag{2.11}$$

Next we define two types of variable order fractional derivatives.

Definition 2.14 (*Left and right Riemann–Liouville derivatives of variable order*) The left Riemann–Liouville derivative of variable fractional order $\alpha(\cdot, \cdot)$ of a function f is defined by

$$\forall t \in (a, b], \quad _a D_t^{\alpha(\cdot,\cdot)}[f](t) := \frac{d}{dt} {}_a I_t^{1-\alpha(\cdot,\cdot)}[f](t),$$

while the right Riemann–Liouville derivative of variable fractional order $\alpha(\cdot, \cdot)$ is defined by

$$\forall t \in [a, b), \quad _t D_b^{\alpha(\cdot,\cdot)}[f](t) := -\frac{d}{dt} {}_t I_b^{1-\alpha(\cdot,\cdot)}[f](t).$$

Definition 2.15 (*Left and right Caputo derivatives of variable fractional order*) The left Caputo derivative of variable fractional order $\alpha(\cdot, \cdot)$ is defined by

$$\forall t \in (a, b], \quad _a^C D_t^{\alpha(\cdot,\cdot)}[f](t) := {}_a I_t^{1-\alpha(\cdot,\cdot)} \left[\frac{d}{dt} f \right](t),$$

while the right Caputo derivative of variable fractional order $\alpha(\cdot, \cdot)$ is given by

$$\forall t \in [a, b), \quad _t^C D_b^{\alpha(\cdot,\cdot)}[f](t) := -{}_t I_b^{1-\alpha(\cdot,\cdot)} \left[\frac{d}{dt} f \right](t).$$

2.1.3 Generalized Fractional Operators

This section presents definitions of one-dimensional generalized fractional operators. In special cases, these operators simplify to the classical Riemann–Liouville
fractional integrals, and Riemann–Liouville and Caputo fractional derivatives. As
before,

$$\Delta := \left\{ (t, \tau) \in \mathbb{R}^2 : a \le \tau < t \le b \right\}.$$

Definition 2.16 (*Generalized fractional integrals of Riemann–Liouville type*) Let
us consider a function k defined almost everywhere on Δ with values in $\mathbb{R}$. For any
function f defined almost everywhere on (a, b) with value in $\mathbb{R}$, the generalized
fractional integral operator K_P is defined for almost all $t \in (a, b)$ by:

$$K_P[f](t) = \lambda \int_a^t k(t, \tau) f(\tau)\, d\tau + \mu \int_t^b k(\tau, t) f(\tau)\, d\tau, \qquad (2.12)$$

with $P = \langle a, t, b, \lambda, \mu \rangle$, $\lambda, \mu \in \mathbb{R}$.

In particular, for suitably chosen kernels $k(t, \tau)$ and sets P, kernel operators K_P
reduce to the classical or variable order fractional integrals of Riemann–Liouville
type, and classical fractional integrals of Hadamard type.

Example 2.17 (a) Let $k^\alpha(t - \tau) = \frac{1}{\Gamma(\alpha)}(t - \tau)^{\alpha-1}$ and $0 < \alpha < 1$. If $P = \langle a, t, b, 1, 0 \rangle$, then

$$K_P[f](t) = \frac{1}{\Gamma(\alpha)} \int_a^t (t - \tau)^{\alpha-1} f(\tau)\, d\tau =: {}_aI_t^\alpha[f](t)$$

is the left Riemann–Liouville fractional integral of order α; if $P = \langle a, t, b, 0, 1 \rangle$,
then

$$K_P[f](t) = \frac{1}{\Gamma(\alpha)} \int_t^b (\tau - t)^{\alpha-1} f(\tau)\, d\tau =: {}_tI_b^\alpha[f](t)$$

is the right Riemann–Liouville fractional integral of order α.
(b) For $k^\alpha(t, \tau) = \frac{1}{\Gamma(\alpha(t,\tau))}(t - \tau)^{\alpha(t,\tau)-1}$ and $P = \langle a, t, b, 1, 0 \rangle$,

$$K_P[f](t) = \int_a^t \frac{1}{\Gamma(\alpha(t, \tau))}(t - \tau)^{\alpha(t,\tau)-1} f(\tau)\, d\tau =: {}_aI_t^{\alpha(\cdot,\cdot)}[f](t)$$

is the left Riemann–Liouville fractional integral of order $\alpha(\cdot, \cdot)$ and for $P = \langle a, t, b, 0, 1 \rangle$

$$K_P[f](t) = \int_t^b \frac{1}{\Gamma(\alpha(\tau,t))} (\tau - t)^{\alpha(t,\tau)-1} f(\tau) \, d\tau =: {}_t I_b^{\alpha(\cdot,\cdot)}[f](t)$$

is the right Riemann–Liouville fractional integral of order $\alpha(\cdot, \cdot)$.

(c) For any $0 < \alpha < 1$, kernel $k^\alpha(t, \tau) = \frac{1}{\Gamma(\alpha)} \left(\log \frac{t}{\tau}\right)^{\alpha-1} \frac{1}{\tau}$ and $P = \langle a, t, b, 1, 0 \rangle$, the general operator K_P reduces to the left Hadamard fractional integral:

$$K_P[f](t) = \frac{1}{\Gamma(\alpha)} \int_a^t \left(\log \frac{t}{\tau}\right)^{\alpha-1} \frac{f(\tau) \, d\tau}{\tau} =: {}_a J_t^\alpha[f](t);$$

and for $P = \langle a, t, b, 0, 1 \rangle$ operator K_P reduces to the right Hadamard fractional integral:

$$K_P[f](t) = \frac{1}{\Gamma(\alpha)} \int_t^b \left(\log \frac{\tau}{t}\right)^{\alpha-1} \frac{f(\tau) \, d\tau}{\tau} =: {}_t J_b^\alpha[f](t).$$

(d) Generalized fractional integrals can be also reduced to, e.g., Riesz, Katugampola or Kilbas fractional operators. Their definitions can be found in Katugampola (2011), Kilbas and Saigo (2004), Kilbas et al. (2006).

The generalized differential operators A_P and B_P are defined with the help of the operator K_P.

Definition 2.18 (*Generalized fractional derivative of Riemann–Liouville type*) The generalized fractional derivative of Riemann–Liouville type, denoted by A_P, is defined by

$$A_P = \frac{d}{dt} \circ K_P.$$

The next differential operator is obtained by interchanging the order of the operators in the composition that defines A_P.

Definition 2.19 (*Generalized fractional derivative of Caputo type*) The general kernel differential operator of Caputo type, denoted by B_P, is given by

$$B_P = K_P \circ \frac{d}{dt}.$$

Example 2.20 The standard Riemann–Liouville and Caputo fractional derivatives (see, e.g., (Kilbas et al. 2006; Klimek 2009; Podlubny 1999; Samko et al. 1993)) are easily obtained from the general kernel operators A_P and B_P, respectively. Let $k^\alpha(t - \tau) = \frac{1}{\Gamma(1-\alpha)} (t - \tau)^{-\alpha}$, $\alpha \in (0, 1)$. If $P = \langle a, t, b, 1, 0 \rangle$, then

$$A_P[f](t) = \frac{1}{\Gamma(1-\alpha)} \frac{d}{dt} \int\limits_a^t (t-\tau)^{-\alpha} f(\tau)\, d\tau =: {}_a D_t^\alpha[f](t)$$

is the standard left Riemann–Liouville fractional derivative of order α, while

$$B_P[f](t) = \frac{1}{\Gamma(1-\alpha)} \int\limits_a^t (t-\tau)^{-\alpha} f'(\tau)\, d\tau =: {}_a^C D_t^\alpha[f](t)$$

is the standard left Caputo fractional derivative of order α; if $P = \langle a, t, b, 0, 1 \rangle$, then

$$-A_P[f](t) = -\frac{1}{\Gamma(1-\alpha)} \frac{d}{dt} \int\limits_t^b (\tau - t)^{-\alpha} f(\tau)\, d\tau =: {}_t D_b^\alpha[f](t)$$

is the standard right Riemann–Liouville fractional derivative of order α, while

$$-B_P[f](t) = -\frac{1}{\Gamma(1-\alpha)} \int\limits_t^b (\tau - t)^{-\alpha} f'(\tau)\, d\tau =: {}_t^C D_b^\alpha[f](t)$$

is the standard right Caputo fractional derivative of order α.

2.2 Multidimensional Fractional Calculus

In this section, we introduce notions of classical, variable order, and generalized partial fractional integrals and derivatives in a multidimensional finite domain. They are natural generalizations of the corresponding fractional operators of Sect. 2.1.1. Furthermore, similarly as in the integer order case, computation of partial fractional derivatives and integrals is reduced to the computation of one-variable fractional operators. Along the work, for $i = 1, \ldots, n$, let a_i, b_i and α_i be numbers in $\mathbb{R}$ and $t = (t_1, \ldots, t_n)$ be such that $t \in \Omega_n$, where $\Omega_n = (a_1, b_1) \times \cdots \times (a_n, b_n)$ is a subset of $\mathbb{R}^n$. Moreover, let us define the following sets: $\Delta_i := \left\{ (t_i, \tau) \in \mathbb{R}^2 : a_i \leq \tau < t_i \leq b_i \right\}, i = 1, \ldots, n$.

2.2.1 Classical Partial Fractional Integrals and Derivatives

In this section we present definitions of classical partial fractional integrals and derivatives. Interested reader can find more details in Sect. 24.1 of the book (Samko et al. 1993).

Definition 2.21 (*Left and right Riemann–Liouville partial fractional integrals*) Let $t \in \Omega_n$. The left and the right partial Riemann–Liouville fractional integrals of order $\alpha_i \in \mathbb{R}$ ($\alpha_i > 0$) with respect to the ith variable t_i are defined by

$$_{a_i} I_{t_i}^{\alpha_i}[f](t) := \frac{1}{\Gamma(\alpha_i)} \int_{a_i}^{t_i} \frac{f(t_1, \ldots, t_{i-1}, \tau, t_{i+1}, \ldots, t_n) \, d\tau}{(t_i - \tau)^{1-\alpha_i}}, \quad t_i > a_i, \quad (2.13)$$

and

$$_{t_i} I_{b_i}^{\alpha_i}[f](t) := \frac{1}{\Gamma(\alpha_i)} \int_{t_i}^{b_i} \frac{f(t_1, \ldots, t_{i-1}, \tau, t_{i+1}, \ldots, t_n) \, d\tau}{(\tau - t_i)^{1-\alpha}}, \quad t_i < b_i, \quad (2.14)$$

respectively.

Definition 2.22 (*Left and right Riemann–Liouville partial fractional derivatives*) Let $t \in \Omega_n$. The left partial Riemann–Liouville fractional derivative of order α_i, $0 < \alpha_i < 1$, of a function f with respect to the ith variable t_i, is defined by $_{a_i} D_{t_i}^{\alpha_i}[f](t) := \frac{\partial}{\partial t_i} {}_{a_i} I_{t_i}^{1-\alpha_i}[f](t)$ for all $t_i \in (a_i, b_i]$. Similarly, the right partial Riemann–Liouville fractional derivative of order α_i of a function f, with respect to the ith variable t_i, is defined by $_{t_i} D_{b_i}^{\alpha_i}[f](t) := -\frac{\partial}{\partial t_i} {}_{t_i} I_{b_i}^{1-\alpha_i}[f](t)$ for all $t_i \in [a_i, b_i)$.

Definition 2.23 (*Left and right Caputo partial fractional derivatives*) Let $t \in \Omega_n$. The left and right partial Caputo fractional derivatives of order α_i, $0 < \alpha_i < 1$, of a function f with respect to the ith variable t_i, are given by

$$_{a_i}^C D_{t_i}^{\alpha_i}[f](t) := {}_{a_i} I_{t_i}^{1-\alpha_i} \left[\frac{\partial}{\partial t_i} f \right](t), \quad \forall t_i \in (a_i, b_i],$$

and

$$_{t_i}^C D_{b_i}^{\alpha_i}[f](t) := -{}_{t_i} I_{b_i}^{1-\alpha_i} \left[\frac{\partial}{\partial t_i} f \right](t), \quad \forall t_i \in [a_i, b_i),$$

respectively.

2.2.2 Variable Order Partial Fractional Integrals and Derivatives

In this section, we introduce the notions of partial fractional operators of variable order. In the following let us assume that $\alpha_i : \Delta_i \to [0, 1]$, $\alpha_i \in C^1\left(\bar{\Delta}; \mathbb{R}\right)$, $i = 1, \ldots, n$, $t \in \Omega_n$ and $f : \Omega_n \to \mathbb{R}$.

Definition 2.24 The left Riemann–Liouville partial integral of variable fractional order $\alpha_i(\cdot, \cdot)$ with respect to the ith variable t_i, is given by

$$_{a_i}I_{t_i}^{\alpha_i(\cdot,\cdot)}[f](t) := \int_{a_i}^{t_i} \frac{1}{\Gamma(\alpha_i(t_i, \tau))}(t_i - \tau)^{\alpha_i(t_i,\tau)-1} f(t_1, \ldots, t_{i-1}, \tau, t_{i+1}, \ldots, t_n) \, d\tau,$$

$t_i > a_i$, while

$$_{t_i}I_{b_i}^{\alpha_i(\cdot,\cdot)}[f](t) := \int_{t_i}^{b_i} \frac{1}{\Gamma(\alpha_i(\tau, t_i))}(\tau - t_i)^{\alpha_i(\tau,t_i)-1} f(t_1, \ldots, t_{i-1}, \tau, t_{i+1}, \ldots, t_n) \, d\tau,$$

$t_i < b_i$, is the right Riemann–Liouville partial integral of variable fractional order $\alpha_i(\cdot, \cdot)$ with respect to variable t_i.

Definition 2.25 The left Riemann–Liouville partial derivative of variable fractional order $\alpha_i(\cdot, \cdot)$, with respect to the ith variable t_i, is given by

$$\forall t_i \in (a_i, b_i], \quad _{a_i}D_{t_i}^{\alpha_i(\cdot,\cdot)}[f](t) = \frac{\partial}{\partial t_i} {}_{a_i}I_{t_i}^{1-\alpha_i(\cdot,\cdot)}[f](t)$$

while the right Riemann–Liouville partial derivative of variable fractional order $\alpha_i(\cdot, \cdot)$, with respect to the ith variable t_i, is defined by

$$\forall t_i \in [a_i, b_i), \quad _{t_i}D_{b_i}^{\alpha_i(\cdot,\cdot)}[f](t) = -\frac{\partial}{\partial t_i} {}_{t_i}I_{b_i}^{1-\alpha_i(\cdot,\cdot)}[f](t).$$

Definition 2.26 The left Caputo partial derivative of variable fractional order $\alpha_i(\cdot, \cdot)$, with respect to the ith variable t_i, is defined by

$$\forall t_i \in (a_i, b_i], \quad _{a_i}^{C}D_{t_i}^{\alpha_i(\cdot,\cdot)}[f](t) = {}_{a_i}I_{t_i}^{1-\alpha_i(\cdot,\cdot)}\left[\frac{\partial}{\partial t_i}f\right](t),$$

while the right Caputo partial derivative of variable fractional order $\alpha_i(\cdot, \cdot)$, with respect to the ith variable t_i, is given by

$$\forall t_i \in [a_i, b_i), \quad _{t_i}^{C}D_{b_i}^{\alpha_i(\cdot,\cdot)}[f](t) = -{}_{t_i}I_{b_i}^{1-\alpha_i(\cdot,\cdot)}\left[\frac{\partial}{\partial t_i}f\right](t).$$

Note that, if $\alpha_i(\cdot, \cdot)$ is a constant function, then the partial operators of variable fractional order are reduced to corresponding partial integrals and derivatives of constant order introduced in Sect. 2.2.1.

2.2.3 Generalized Partial Fractional Operators

Let us assume that $\lambda = (\lambda_1, \ldots, \lambda_n)$ and $\mu = (\mu_1, \ldots, \mu_n)$ are in $\mathbb{R}^n$. We shall present definitions of generalized partial fractional integrals and derivatives. Let $k_i : \Delta_i \to \mathbb{R}, i = 1, \ldots, n$ and $t \in \Omega_n$.

Definition 2.27 (*Generalized partial fractional integral*) For any function f defined almost everywhere on Ω_n with value in $\mathbb{R}$, the generalized partial integral K_{P_i} is defined for almost all $t_i \in (a_i, b_i)$ by:

$$K_{P_i}[f](t) := \lambda_i \int_{a_i}^{t_i} k_i(t_i, \tau) f(t_1, \ldots, t_{i-1}, \tau, t_{i+1}, \ldots, t_n) \, d\tau$$

$$+ \mu_i \int_{t_i}^{b_i} k_i(\tau, t_i) f(t_1, \ldots, t_{i-1}, \tau, t_{i+1}, \ldots, t_n) \, d\tau,$$

where $P_i = \langle a_i, t_i, b_i, \lambda_i, \mu_i \rangle$.

Definition 2.28 (*Generalized partial fractional derivative of Riemann–Liouville type*) The generalized partial fractional derivative of Riemann–Liouville type with respect to the ith variable t_i is given by $A_{P_i} := \frac{\partial}{\partial t_i} \circ K_{P_i}$.

Definition 2.29 (*Generalized partial fractional derivative of Caputo type*) The generalized partial fractional derivative of Caputo type with respect to the ith variable t_i is given by $B_{P_i} := K_{P_i} \circ \frac{\partial}{\partial t_i}$.

Example 2.30 Similarly, as in the one-dimensional case, partial operators K, A and B reduce to the standard partial fractional integrals and derivatives. The left- or right-sided Riemann–Liouville partial fractional integral with respect to the ith variable t_i is obtained by choosing the kernel $k_i^\alpha(t_i, \tau) = \frac{1}{\Gamma(\alpha_i)}(t_i - \tau)^{\alpha_i - 1}$. That is,

$$K_{P_i}[f](t) = \frac{1}{\Gamma(\alpha_i)} \int_{a_i}^{t_i} (t_i - \tau)^{\alpha_i - 1} f(t_1, \ldots, t_{i-1}, \tau, t_{i+1}, \ldots, t_n) \, d\tau =: {_{a_i}}I_{t_i}^{\alpha_i}[f](t),$$

for $P_i = \langle a_i, t_i, b_i, 1, 0 \rangle$, and

$$K_{P_i}[f](t) = \frac{1}{\Gamma(\alpha_i)} \int_{t_i}^{b_i} (\tau - t_i)^{\alpha_i - 1} f(t_1, \ldots, t_{i-1}, \tau, t_{i+1}, \ldots, t_n) \, d\tau =: {_{t_i}}I_{b_i}^{\alpha_i}[f](t),$$

for $P_i = \langle a_i, t_i, b_i, 0, 1 \rangle$. The standard left- and right-sided Riemann–Liouville and Caputo partial fractional derivatives with respect to ith variable t_i are received by choosing the kernel $k_i^\alpha(t_i, \tau) = \frac{1}{\Gamma(1-\alpha_i)}(t_i - \tau)^{-\alpha_i}$. If $P_i = \langle a_i, t_i, b_i, 1, 0 \rangle$, then

$$A_{P_i}[f](t) = \frac{1}{\Gamma(1 - \alpha_i)} \frac{\partial}{\partial t_i} \int_{a_i}^{t_i} (t_i - \tau)^{-\alpha_i} f(t_1, \ldots, t_{i-1}, \tau, t_{i+1}, \ldots, t_n) \, d\tau$$

$$=: {}_{a_i}D_{t_i}^{\alpha_i}[f](t),$$

$$B_{P_i}[f](t) = \frac{1}{\Gamma(1 - \alpha_i)} \int_{a_i}^{t_i} (t_i - \tau)^{-\alpha_i} \frac{\partial}{\partial \tau} f(t_1, \ldots, t_{i-1}, \tau, t_{i+1}, \ldots, t_n) \, d\tau$$

$$=: {}_{a_i}^{C}D_{t_i}^{\alpha_i}[f](t).$$

If $P_i = \langle a_i, t_i, b_i, 0, 1 \rangle$, then

$$-A_{P_i}[f](t) = \frac{-1}{\Gamma(1 - \alpha_i)} \frac{\partial}{\partial t_i} \int_{t_i}^{b_i} (\tau - t_i)^{-\alpha_i} f(t_1, \ldots, t_{i-1}, \tau, t_{i+1}, \ldots, t_n) \, d\tau$$

$$=: {}_{t_i}D_{b_i}^{\alpha_i}[f](t),$$

$$-B_{P_i}[f](t) = \frac{-1}{\Gamma(1 - \alpha_i)} \int_{t_i}^{b_i} (\tau - t_i)^{-\alpha_i} \frac{\partial}{\partial \tau} f(t_1, \ldots, t_{i-1}, \tau, t_{i+1}, \ldots, t_n) \, d\tau$$

$$=: {}_{t_i}^{C}D_{b_i}^{\alpha_i}[f](t).$$

Moreover, one can easily check that also variable order partial fractional integrals and derivatievs are particular cases of operators K_{P_i}, A_{P_i} and B_{P_i}.

References

Agrawal OP (2010) Generalized variational problems and Euler-Lagrange equations. Comput Math Appl 59(5):1852–1864

Almeida A, Samko S (2009) Fractional and hypersingular operators in variable exponent spaces on metric measure spaces. Mediterr J Math 6:215–232

Bourdin L, Odzijewicz T, Torres DFM (2014) Existence of minimizers for generalized Lagrangian functionals and a necessary optimality condition—application to fractional variational problems. Differ Integral Equ 27(7–8):743–766

Coimbra CFM (2003) Mechanics with variable-order differential operators. Ann Phys 12(11–12): 692–703

Cresson J (2007) Fractional embedding of differential operators and Lagrangian systems. J Math Phys 48(3):033504, 34 pp

Diaz G, Coimbra CFM (2009) Nonlinear dynamics and control of a variable order oscillator with application to the van der Pol equation. Nonlinear Dyn 56(1–2):145–157

Hilfer R (2000) Applications of fractional calculus in physics. World Scientific Publishing, River Edge

Katugampola UN (2011) New approach to a generalized fractional integral. Appl Math Comput 218(3):860–865

Kilbas AA, Saigo M (2004) Generalized Mittag-Leffler function and generalized fractional calculus operators. Integral Transform Spec Func 15(1):31–49

Kilbas AA, Srivastava HM, Trujillo JJ (2006) Theory and applications of fractional differential equations, vol 204. North-Holland mathematics studies. Elsevier, Amsterdam

Klimek M (2005) Lagrangian fractional mechanics—a noncommutative approach. Czechoslovak J Phys 55(11):1447–1453

Klimek M (2009) On solutions of linear fractional differential equations of a variational type. The Publishing Office of Czestochowa University of Technology, Czestochowa

Klimek M, Lupa M (2013) Reflection symmetric formulation of generalized fractional variational calculus. Fract Calc Appl Anal 16(1):243–261

Lacroix SF (1819) Traite du calcul differentiel et du calcul integral. Paris: Mme. VeCourcier, second edition 3:409–410

Lorenzo CF, Hartley TT (2002) Variable order and distributed order fractional operators. Nonlinear Dyn 29(1–4):57–98

Odzijewicz T, Malinowska AB, Torres DFM (2012a) Generalized fractional calculus with applications to the calculus of variations. Comput Math Appl 64(10):3351–3366

Odzijewicz T, Malinowska AB, Torres DFM (2012b) Fractional calculus of variations in terms of a generalized fractional integral with applications to physics. Abstr Appl Anal 2012(871912), 24 pp

Odzijewicz T, Malinowska AB, Torres DFM (2013a) Green's theorem for generalized fractional derivative. Fract Calc Appl Anal 16(1):64–75

Odzijewicz T, Malinowska AB, Torres DFM (2013b) A generalized fractional calculus of variations. Control Cybern 42(2):443–458

Odzijewicz T, Malinowska AB, Torres DFM (2013c) Fractional calculus of variations of several independent variables. Eur Phys J Spec Top 222(8):1813–1826

Pedro HTC, Kobayashi MH, Pereira JMC, Coimbra CFM (2008) Variable order modeling of diffusive-convective effects on the oscillatory flow past a sphere. J Vib Control 14(9–10):1569–1672

Podlubny I (1999) Fractional differential equations. Academic Press, San Diego

Ramirez LES, Coimbra CFM (2010) On the selection and meaning of variable order operators for dynamic modeling. Int J Differ Equ 2010(846107):16

Ramirez LES, Coimbra CFM (2011) On the variable order dynamics of the nonlinear wake caused by a sedimenting particle. Phys D 240(13):1111–1118

Samko SG, Kilbas AA, Marichev OI (1993) Fractional integrals and derivatives. Translated from the 1987 Russian original. Gordon and Breach, Yverdon

Samko SG, Ross B (1993) Integration and differentiation to a variable fractional order. Integral Transform Spec Funct 1(4):277–300

Chapter 3
Fractional Calculus of Variations

Abstract We review a few main approaches to the fractional calculus of variations.

Keywords Calculus of variations · Fractional calculus of variations · Lagrangian · Fractional Euler–Lagrange equations · Fractional embedding · Coherence problem

The calculus of variations is a beautiful and useful field of mathematics that deals with the problems of determining extrema (maxima or minima) of functionals (Dacorogna 2004; Malinowska and Torres 2012, 2014). For the first time, serious attention of scientists was directed to the variational calculus in 1696, when Johann Bernoulli asked about the curve with specified endpoints, lying in a vertical plane, for which the time taken by a material point sliding without friction and under gravity from one end to the other is minimal. This problem gained interest of such scientists as Leibniz, Newton, or L'Hôpital and was called the brachistochrone problem. Afterward, a student of Bernoulli, the brilliant Swiss mathematician Leonhard Euler, considered the problem of finding a function extremizing (minimizing or maximizing) an integral

$$\mathcal{J}(y) = \int_a^b F(y(t), \dot{y}(t), t)\, dt \qquad (3.1)$$

subject to the boundary conditions

$$y(a) = y_a \quad \text{and} \quad y(b) = y_b \qquad (3.2)$$

with $y \in C^2([a, b]; \mathbb{R})$, $a, b, y_a, y_b \in \mathbb{R}$ and $F(u, v, t)$ satisfying some smoothness properties. He proved that the curve $y(t)$ must satisfy the following necessary condition, so-called Euler–Lagrange equation:

$$\frac{\partial F(y(t), \dot{y}(t), t)}{\partial u} - \frac{d}{dt}\left(\frac{\partial F(y(t), \dot{y}(t), t)}{\partial v}\right) = 0. \qquad (3.3)$$

Solutions of Eq. (3.3) are usually called extremals. It is important to remark that the calculus of variations is a very interesting topic because of its numerous applications

© The Author(s) 2015

A.B. Malinowska et al., *Advanced Methods in the Fractional Calculus of Variations*, SpringerBriefs in Applied Sciences and Technology, DOI 10.1007/978-3-319-14756-7_3

in geometry and differential equations, in mechanics and physics, and in areas as diverse as engineering, medicine, economics, and renewable resources (Clarke 2013).

In the next example, we give a simple application of the calculus of variations. Precisely, we present the soap bubble problem, stated by Euler in 1744.

Example 3.1 (cf. Example 14.1 (Clarke 2013)) In the soap bubble problem we want to find a surface of rotation, spanned by two concentric rings of radius A and B, which has the minimum area. This wish is confirmed by experiment and is based on d'Alembert principle. In the sense of the calculus of variations, we can formulate the soap bubble problem in the following way: we want to minimize the variational functional

$$\mathcal{J}(y) = \int_a^b y(t)\sqrt{1 + \dot{y}(t)^2}\, dt \quad \text{subject to} \quad y(a) = A, \quad y(b) = B.$$

This is a special case of problem (3.1) and (3.2) with $F(u, v, t) = u\sqrt{1 + v^2}$. Let $y(t) > 0$, $\forall t$. It is not difficult to verify that the Euler–Lagrange equation is given by

$$\ddot{y}(t) = \frac{1 + \dot{y}(t)^2}{y(t)}$$

and its solution is the catenary curve given by

$$y(t) = k \cosh\left(\frac{t + c}{k}\right),$$

where c, k are certain constants.

This book is devoted to the fractional calculus of variations and its generalizations. Therefore, in the next sections, we present the basic results of the noninteger variational calculus. Let us precise, that along the work we will understand $\partial_i F$ as the partial derivative of function F with respect to its ith argument.

3.1 Fractional Euler–Lagrange Equations

Within the years, several methods were proposed to solve mechanical problems with nonconservative forces, e.g., Rayleigh dissipation function method, a technique introducing an auxiliary coordinate or approaches including the microscopic details of the dissipation directly in the Lagrangian. Although all mentioned methods are correct, they are not as direct and simple as it is in the case of conservative systems. In his notes from 1996–1997, Riewe presented a new approach to nonconservative forces (Riewe 1996, 1997). He claimed that friction forces follow from Lagrangians containing terms proportional to fractional derivatives. Precisely, for $y : [a, b] \to \mathbb{R}^r$ and $\alpha_i, \beta_j \in [0, 1]$, $i = 1, \ldots, N$, $j = 1, \ldots, N'$, he considered the following energy functional:

$$\mathcal{J}(y) = \int\limits_a^b F\left({}_aD_t^{\alpha_1}[y](t), \ldots, {}_aD_t^{\alpha_N}[y](t), {}_tD_b^{\beta_1}[y](t), \ldots, {}_tD_b^{\beta_{N'}}[y](t), y(t), t\right) dt,$$

with r, N, and N' being natural numbers. Using the fractional variational principle he obtained the following Euler–Lagrange equation:

$$\sum_{i=1}^{N} {}_tD_b^{\alpha_i}[\partial_i F] + \sum_{i=1}^{N'} {}_aD_t^{\beta_i}[\partial_{i+N} F] + \partial_{N'+N+1} F = 0. \tag{3.4}$$

Riewe illustrated his results through the classical problem of linear friction.

Example 3.2 (Riewe 1997) Let us consider the following Lagrangian:

$$F = \frac{1}{2}m\dot{y}^2 - V(y) + \frac{1}{2}\gamma i \left({}_aD_t^{\frac{1}{2}}[y]\right)^2, \tag{3.5}$$

where the first term in the sum represents kinetic energy, the second one represents potential energy, the last one is linear friction energy, and $i^2 = -1$. Using (3.4), we can obtain the Euler–Lagrange equation for a Lagrangian containing derivative of order one and order $\frac{1}{2}$:

$$\frac{\partial F}{\partial y} + {}_tD_b^{\frac{1}{2}}\left[\frac{\partial F}{\partial\, {}_aD_t^{\frac{1}{2}}[y]}\right] - \frac{d}{dt}\frac{\partial F}{\partial \dot{y}} = 0,$$

which, in the case of Lagrangian (3.5), becomes

$$m\ddot{y} = -\gamma i \left({}_tD_b^{\frac{1}{2}} \circ {}_aD_t^{\frac{1}{2}}\right)[y] - \frac{\partial V(y)}{\partial y}.$$

In order to obtain the equation with linear friction, $m\ddot{y} + \gamma\dot{y} + \frac{\partial V}{\partial y} = 0$, Riewe suggested considering an infinitesimal time interval, that is, the limiting case $a \to b$, while keeping $a < b$.

After the works of Riewe several authors contributed to the theory of the fractional variational calculus. First, let us point out the approach discussed by Klimek (2005) (see also Klimek (2001, 2002)). It was suggested to study symmetric fractional derivatives of order α, $0 < \alpha < 1$, defined as follows:

$$\mathcal{D}^\alpha := \frac{1}{2}{}_aD_t^\alpha + \frac{1}{2}{}_tD_b^\alpha.$$

In contrast to the left and right fractional derivatives, operator $\mathcal{D}^\alpha$ is symmetric for the scalar product given by

$$\langle f | g \rangle := \int_a^b \overline{f(t)} g(t) \, dt,$$

that is,

$$\langle \mathcal{D}^\alpha [f] | g \rangle = \langle f | \mathcal{D}^\alpha [g] \rangle,$$

where the $\overline{f(t)}$ denotes the complex conjugate of $f(t)$. With this notion for the fractional derivative, for $\alpha_i \in (0, 1)$ and $y : [a, b] \to \mathbb{R}^r$, $i = 1, \ldots, N$, Klimek considered the following action functional:

$$\mathcal{J}(y) = \int_a^b F \left(\mathcal{D}^{\alpha_1}[y](t), \ldots, \mathcal{D}^{\alpha_N}[y](t), y(t), t \right) dt. \qquad (3.6)$$

Using the fractional variational principle, she derived the Euler–Lagrange equation given by

$$\partial_{N+1} F + \sum_{i=1}^N \mathcal{D}^{\alpha_i} [\partial_i F] = 0. \qquad (3.7)$$

As an example, Klimek considered the variational functional

$$\mathcal{J}(y) = \int_a^b 2m\dot{y}^2(t) - \gamma i \left(\mathcal{D}^{\frac{1}{2}}[y](t) \right)^2 - V(y(t)) \, dt$$

and, under appropriate assumptions, she arrived to the equation with linear friction:

$$m\ddot{y} = -\frac{\partial V}{\partial y} - \gamma \dot{y}. \qquad (3.8)$$

Another type of problems, containing Riemann–Liouville fractional derivatives, was discussed by Klimek (2009):

$$\mathcal{J}(y) = \int_a^b F({}_a D_t^{\alpha_1}[y](t), \ldots, {}_a D_t^{\alpha_N}[y](t), y(t), t) \, dt$$

and the Euler–Lagrange equation

$$\partial_{N+1} F + \sum_{i=1}^N {}_t^C D_b^{\alpha_i} [\partial_i F] = 0, \qquad (3.9)$$

containing fractional derivatives of Caputo type, was obtained.

The next examples are borrowed from Klimek (2009).

Example 3.3 (cf. Example 4.1.1 of (Klimek 2009)) Let $0 < \alpha < 1$ and y be a minimizer of the functional

$$\mathcal{J}(y) = \int\limits_a^b \frac{1}{2} y(t) {}_a D_t^\alpha [y](t) \, dt.$$

Then y is a solution to the following Euler–Lagrange equation:

$$\frac{1}{2} \left({}_a D_t^\alpha [y] + {}_t^C D_b^\alpha [y] \right) = 0.$$

Example 3.4 (cf. Example 4.1.2 of (Klimek 2009)) Let $0 < \alpha < 1$. The model of harmonic oscillator, in the framework of classical mechanics, is connected to an action

$$\mathcal{J}(y) = \int\limits_a^b \left[-\frac{1}{2} y'^2(t) + \frac{\omega^2}{2} y^2(t) \right] dt, \tag{3.10}$$

and is determined by the equation

$$y'' + \omega^2 y = 0. \tag{3.11}$$

If in functional (3.10) instead of the derivative of order one we put a derivative of fractional order α, then

$$\mathcal{J}(y) = \int\limits_a^b \left[-\frac{1}{2} \left({}_a D_t^\alpha [y](t) \right)^2 + \frac{\omega^2}{2} y^2(t) \right] dt$$

and by (3.9) the Euler–Lagrange equation has the following form:

$$- {}_t^C D_b^\alpha \left[{}_a D_t^\alpha [y] \right] + \omega^2 y = 0. \tag{3.12}$$

If $\alpha \to 1^+$, then Eq. (3.12) reduces to (3.11). The proof of this fact, as well as solutions to fractional harmonic oscillator equation (3.12), can be found in Klimek (2009).

3.2 Fractional Embedding of Euler–Lagrange Equations

The notion of embedding introduced in Cresson and Darses (2007) is an algebraic procedure providing an extension of classical differential equations over an arbitrary vector space. This formalism is developed in the framework of stochastic processes

(Cresson and Darses 2007), nondifferentiable functions (Cresson et al. 2009), and fractional equations (Cresson 2007). The general scheme of embedding theories is the following: (i) fix a vector space V and a mapping $\iota : C^0([a, b], \mathbb{R}^n) \to V$; (ii) extend differential operators over V; and (iii) extend the notion of integral over V. Let (ι, D, J) be a given embedding formalism, where a linear operator $D : V \to V$ takes place for a generalized derivative on V, and a linear operator $J : V \to \mathbb{R}$ takes place for a generalized integral on V. The embedding procedure gives two different ways, a priori, to generalize Euler–Lagrange equations. The first (pure algebraic) way is to make a direct embedding of the Euler–Lagrange equation. The second (analytic) is to embed the Lagrangian functional associated to the equation and to derive, by the associated calculus of variations, the Euler–Lagrange equation for the embedded functional. A natural question is then the problem of coherence between these two extensions:

Coherence Problem. *Let (ι, D, J) be a given embedding formalism. Do we have equivalence between the Euler–Lagrange equation which gives the direct embedding and the one received from the embedded Lagrangian system?*

As shown in the work (Cresson 2007) for standard fractional differential calculus, the answer to the question above is known to be negative. To be more precise, let us define the following operator first introduced in Cresson (2007).

Definition 3.5 (*Fractional operator of order* (α, β)) Let $a, b \in \mathbb{R}, a < b$ and $\mu \in \mathbb{C}$. We define the fractional operator of order (α, β), with $\alpha > 0$ and $\beta > 0$, by

$$\mathcal{D}_{\mu}^{\alpha,\beta} = \frac{1}{2}\left[{}_aD_t^{\alpha} - {}_tD_b^{\beta}\right] + \frac{i\mu}{2}\left[{}_aD_t^{\alpha} + {}_tD_b^{\beta}\right]. \tag{3.13}$$

In particular, for $\alpha = \beta = 1$ one has $\mathcal{D}_{\mu}^{1,1} = \frac{d}{dt}$. Moreover, for $\mu = -i$ we recover the left Riemann–Liouville fractional derivative of order α,

$$\mathcal{D}_{-i}^{\alpha,\beta} = {}_aD_t^{\alpha},$$

and for $\mu = i$ the right Riemann–Liouville fractional derivative of order β:

$$\mathcal{D}_{i}^{\alpha,\beta} = -{}_tD_b^{\alpha}.$$

Now, let us consider the following variational functional:

$$\mathcal{J}(y) = \int_a^b F(\mathcal{D}_{\mu}^{\alpha,\beta}[y](t), y(t), t)\,dt$$

defined on the space of continuous functions such that ${}_aD_t^{\alpha}[y]$ together with ${}_tD_b^{\beta}[y]$ exist and $y(a) = y_a, y(b) = y_b$. Using the direct embedding procedure, the Euler–Lagrange equation derived by Cresson is

$$\mathcal{D}_{\mu}^{\beta,\alpha}[\partial_1 F] = \partial_2 F. \tag{3.14}$$

Using the fractional variational principle in derivation of the Euler–Lagrange equation, one has

$$\mathcal{D}_{-\mu}^{\beta,\alpha}[\partial_1 F] = \partial_2 F. \tag{3.15}$$

Readers can easily notice that, in general, there is a difference between Eqs. (3.14) and (3.15), i.e., they are not coherent. Cresson claimed (Cresson 2007) that this lack of coherence has the following sources:

- the set of variations in the method of variational principle is to large and therefore it does not give correct answer; one should find the corresponding constraints for the variations;
- there is a relation between lack of coherence and properties of the operator used to generalize the classical derivative.

Let us observe that coherence between (3.14) and (3.15) is restored in the case when $\alpha = \beta$ and $\mu = 0$. This type of coherence is called time reversible coherence. For a deeper discussion of the subject, we refer the reader to Cresson (2007).

In this chapter, we presented few results of the fractional calculus of variations. A comprehensive study of the subject can be found in the books (Almeida et al. 2015; Klimek 2009; Malinowska and Torres 2012).

References

Almeida R, Pooseh S, Torres DFM (2015) Computational methods in the fractional calculus of variations. Imperial College Press, London
Clarke F (2013) Functional analysis, calculus of variations and optimal control. Graduate texts in mathematics. Springer, London
Cresson J (2007) Fractional embedding of differential operators and Lagrangian systems. J Math Phys 48(3):033504, 34 pp
Cresson J, Darses S (2007) Stochastic embedding of dynamical systems. J Math Phys 48(7):072703, 54 pp
Cresson J, Frederico GSF, Torres DFM (2009) Constants of motion for non-differentiable quantum variational problems. Topol Methods Nonlinear Anal 33(2):217–231
Dacorogna B (2004) Introduction to the calculus of variations. Translated from the 1992 French original. Imperial College Press, London
Klimek M (2001) Fractional sequential mechanics—model with symmetric fractional derivatives. Czech J Phys 51(12):1348–1354
Klimek M (2002) Lagrangian and Hamiltonian fractional sequential mechanics. Czech J Phys 52(11):1247–1253
Klimek M (2005) Lagrangian fractional mechanics—a noncommutative approach. Czechoslov J Phys 55(11):1447–1453
Klimek M (2009) On solutions of linear fractional differential equations of a variational type. The Publishing Office of Czestochowa University of Technology, Czestochowa
Malinowska AB, Torres DFM (2012) Introduction to the fractional calculus of variations. Imperial College Press, London

Malinowska AB, Torres DFM (2014) Quantum variational calculus. Springer briefs in electrical and computer engineering. Springer, Cham

Riewe F (1996) Nonconservative Lagrangian and Hamiltonian mechanics. Phys Rev E (3) 53(2):1890–1899

Riewe F (1997) Mechanics with fractional derivatives. Phys Rev E (3), part B 55(3):3581–3592

Chapter 4
Standard Methods in Fractional Variational Calculus

Abstract We investigate the problem of finding an admissible function giving a minimum value to an integral functional that depends on an unknown function (or functions) of one or several variables and its generalized fractional derivatives and/or generalized fractional integrals. The appropriate Euler–Lagrange equations and natural boundary conditions are obtained. Moreover, Noether-type theorems (without transformation of time) are presented.

Keywords Generalized fractional calculus of variations · Fractional integration by parts · Isoperimetric problems · Natural boundary conditions · Fractional Euler–Lagrange equations · Fractional Noether's theorem

We make use of standard methods in the fractional calculus of variations (see, e.g., (Malinowska and Torres 2012)). Namely, by analogy to the classical variational calculus (see, e.g., (Dacorogna 2004)), the approach that we call standard, is first to prove Euler–Lagrange equations, find their solutions, and then to check if they are minimizers (or maximizers). It is important to remark that standard methods suffer an important disadvantage. Precisely, solvability of Euler–Lagrange equations is assumed, which is not the case in direct methods that are going to be presented later (see Chap. 5). The results of this chapter were published in Odzijewicz (2013), Odzijewicz et al. (2012a, b, c, d, e, 2013a, b, c), Odzijewicz and Torres (2011, 2012).

Before we describe briefly an arrangement of this chapter, we define the concept of minimizer. Let $(X, \|\cdot\|)$ be a normed linear space and $\mathcal{I}$ be a functional defined on a nonempty subset $\mathcal{A}$ of X. Moreover, let us introduce the following set: if $\bar{y} \in \mathcal{A}$ and $\delta > 0$, then

$$\mathcal{N}_\delta(\bar{y}) := \{y \in \mathcal{A} : \|y - \bar{y}\| < \delta\}$$

is called neighborhood of $\bar{y}$ in $\mathcal{A}$.

Definition 4.1 Function $\bar{y} \in \mathcal{A}$ is called minimizer of $\mathcal{I}$ if there exists a neighborhood $\mathcal{N}_\delta(\bar{y})$ of $\bar{y}$ such that

$$\mathcal{I}(\bar{y}) \leq \mathcal{I}(y), \quad \text{for all } y \in \mathcal{N}_\delta(\bar{y}).$$

© The Author(s) 2015

A.B. Malinowska et al., *Advanced Methods in the Fractional Calculus of Variations*, SpringerBriefs in Applied Sciences and Technology, DOI 10.1007/978-3-319-14756-7_4

Note that any function $y \in \mathcal{N}_\delta(\bar{y})$ can be represented in a convenient way as a perturbation of $\bar{y}$. Precisely,

$$\forall y \in \mathcal{N}_\delta(\bar{y}), \quad \exists \eta \in \mathcal{A}_0, \quad y = \bar{y} + h\eta, \quad |h| \le \varepsilon,$$

where $0 < \varepsilon < \frac{\delta}{\|\eta\|}$ and $\mathcal{A}_0$ is a suitable set of functions η such that

$$\mathcal{A}_0 = \{\eta \in X : \bar{y} + h\eta \in \mathcal{A}, \quad |h| \le \varepsilon\}.$$

We begin the chapter with Sect. 4.1, where we prove generalized integration by parts formula and boundedness of generalized fractional integral from $L^p(a, b; \mathbb{R})$ to $L^q(a, b; \mathbb{R})$.

In Sect. 4.2, we consider the one-dimensional fundamental problem with generalized fractional operators and obtain an appropriate Euler–Lagrange equation. Then, we prove that under some convexity assumptions on Lagrangian, every solution to the Euler–Lagrange equation is automatically a solution to our problem. Moreover, as corollaries, we obtain results for problems of the constant and variable order fractional variational calculus and discuss some illustrative examples.

In Sect. 4.3, we study variational problems with free end points and besides Euler–Lagrange equations we prove natural boundary conditions. As particular cases we obtain natural boundary conditions for problems with constant and variable order fractional operators.

Section 4.4 is devoted to generalized fractional isoperimetric problems. We want to find functions that minimize an integral functional subject to given boundary conditions and isoperimetric constraints. We prove necessary optimality conditions and, as corollaries, we obtain Euler–Lagrange equations for isoperimetric problems with constant and variable order fractional operators. Furthermore, we illustrate our results through several examples.

In Sect. 4.5, we prove a generalized fractional counterpart of Noether's theorem without transformation of time. Assuming invariance of the functional, we prove that any extremal must satisfy a certain generalized fractional equation. Corresponding results are obtained for functionals with constant and variable order fractional operators.

Section 4.6 is dedicated to variational problems defined by the use of the generalized fractional integral instead of the classical integral. We obtain Euler–Lagrange equations and discuss several examples.

Finally, in Sect. 4.7 we study multidimensional fractional variational problems with generalized partial operators. We begin with the proofs of integration by parts formulas for generalized partial fractional integrals and derivatives. Next, we use these results to show Euler–Lagrange equations for the fundamental problem. Moreover, we prove a generalized fractional Dirichlet's principle, a necessary optimality condition for the isoperimetric problem and Noether's theorem. We finish the chapter with some conclusions.

4.1 Properties of Generalized Fractional Integrals

This section is devoted to properties of generalized fractional operators. We begin by proving in Sect. 4.1.1 that the generalized fractional operator K_P is bounded and linear. Later, in Sect. 4.1.2, we give integration by parts formulas for generalized fractional integrals.

4.1.1 Boundedness of Generalized Fractional Operators

Along the work, we assume that $1 < p < \infty$ and that q is an adjoint of p, that is, $\frac{1}{p} + \frac{1}{q} = 1$. Let us prove the following theorem yielding boundedness of the generalized fractional integral K_P from $L^p(a, b; \mathbb{R})$ to $L^q(a, b; \mathbb{R})$.

Theorem 4.2 *Let us assume that $k \in L^q(\Delta; \mathbb{R})$. Then, K_P is a linear bounded operator from $L^p(a, b; \mathbb{R})$ to $L^q(a, b; \mathbb{R})$.*

Proof The linearity is obvious. We will show that K_P is bounded from $L^p(a, b; \mathbb{R})$ to $L^q(a, b; \mathbb{R})$. Considering only the first term of K_P, let us prove that the following inequality holds for any $f \in L^p(a, b; \mathbb{R})$:

$$\left(\int_a^b \left| \int_a^t k(t, \tau) f(\tau) \, d\tau \right|^q dt \right)^{1/q} \leq \|k\|_{L^q(\Delta, \mathbb{R})} \|f\|_{L^p}. \tag{4.1}$$

Using Fubini's theorem, we have $k(t, \cdot) \in L^q(a, t; \mathbb{R})$ for almost all $t \in (a, b)$. Then, applying Hölder's inequality, we have

$$\left| \int_a^t k(t, \tau) f(\tau) \, d\tau \right|^q \leq \left[\left(\int_a^t |k(t, \tau)|^q \, d\tau \right)^{\frac{1}{q}} \left(\int_a^t |f(\tau)|^p \, d\tau \right)^{\frac{1}{p}} \right]^q$$

$$\leq \int_a^t |k(t, \tau)|^q \, d\tau \, \|f\|_{L^p}^q \tag{4.2}$$

for almost all $t \in (a, b)$. Hence, integrating Eq. (4.2) on the interval (a, b), we obtain inequality (4.1). The proof is completed using the same strategy on the second term in the definition of K_P.

Corollary 4.3 *If $\frac{1}{p} < \alpha < 1$, then $_aI_t^\alpha$ is a linear bounded operator from $L^p(a, b; \mathbb{R})$ to $L^q(a, b; \mathbb{R})$.*

Proof Let us denote $k^\alpha(t, \tau) = \frac{1}{\Gamma(\alpha)}(t - \tau)^{\alpha-1}$. For $\frac{1}{p} < \alpha < 1$ there exists a constant $C \in \mathbb{R}$ such that for almost all $t \in (a, b)$

$$\int_a^t |k^\alpha(t, \tau)|^q \, d\tau \le C. \tag{4.3}$$

Integrating (4.3) on (a, b) we have $k^\alpha(t, \tau) \in L^q(\Delta; \mathbb{R})$. Therefore, applying Theorem 4.2 with $P = \langle a, t, b, 1, 0 \rangle$ operator $_aI_t^\alpha$ is linear bounded from $L^P(a, b; \mathbb{R})$ to $L^q(a, b; \mathbb{R})$.

Next result shows that with the use of Theorem 4.2, one can prove that the variable order fractional integral is a linear bounded operator.

Corollary 4.4 *Let* $\alpha : \Delta \rightarrow [\delta, 1]$ *with* $\delta > \frac{1}{p}$. *Then* $_aI_t^{\alpha(\cdot,\cdot)}$ *is a linear bounded operator from* $L^P(a, b; \mathbb{R})$ *to* $L^q(a, b; \mathbb{R})$.

Proof Let us denote $k^\alpha(t, \tau) = (t - \tau)^{\alpha(t,\tau)-1}/\Gamma(\alpha(t, \tau))$. We just have to prove that $k^\alpha \in L^q(\Delta, \mathbb{R})$ in order to use Theorem 4.2. Let us note that since α is with values in $[\delta, 1]$ with $\delta > 0$, then $1/(\Gamma \circ \alpha)$ is bounded. Hence, we have just to prove that $(\Gamma \circ \alpha)k^\alpha \in L^q(\Delta, \mathbb{R})$. We have two different cases: $b - a \le 1$ and $b - a > 1$.

In the first case, for any $(t, \tau) \in \Delta$, we have $0 < t - \tau \le 1$ and $q(\delta - 1) > -1$. Then:

$$\int_a^t (t - \tau)^{q(\alpha(t,\tau)-1)} \, d\tau \le \int_a^t (t - \tau)^{q(\delta-1)} \, d\tau = \frac{(t - a)^{q(\delta-1)+1}}{q(\delta - 1) + 1} \le \frac{1}{q(\delta - 1) + 1}.$$

In the second case, for almost all $(t, \tau) \in \Delta \cap (a, a + 1) \times (a, b)$, we have $0 < t - \tau \le 1$. Consequently, we conclude in the same way that:

$$\int_a^t (t - \tau)^{q(\alpha(t,\tau)-1)} \, d\tau \le \frac{1}{q(\delta - 1) + 1}.$$

Still in the second case, for almost all $(t, \tau) \in \Delta \cap (a + 1, b) \times (a, b)$, we have $\tau < t - 1$ or $t - 1 \le \tau \le t$. Then:

$$\int_a^t (t - \tau)^{q(\alpha(t,\tau)-1)} \, d\tau = \int_a^{t-1} (t - \tau)^{q(\alpha(t,\tau)-1)} \, d\tau + \int_{t-1}^t (t - \tau)^{q(\alpha(t,\tau)-1)} \, d\tau$$

$$\le b - a - 1 + \frac{1}{q(\delta - 1) + 1}.$$

Consequently, in any case, there exists a constant $C \in \mathbb{R}$ such that for almost all $t \in (a, b)$:

$$\int\limits_{a}^{t} \left| k^{\alpha}(t, \tau) \right|^{q} \, d\tau \leq C. \tag{4.4}$$

Finally, $k^{\alpha} \in L^{q}(\Delta, \mathbb{R})$.

4.1.2 Generalized Fractional Integration by Parts

In this section we obtain a formula of integration by parts for the generalized fractional calculus. Our results are particularly useful with respect to applications in dynamic optimization, where the derivation of the Euler–Lagrange equations uses, as a key step in the proof, integration by parts (see, e.g., the proof of Theorem 4.10 in Sect. 4.2).

In our setting, integration by parts changes a given parameter set P into its dual P^{*}. The term *duality* comes from the fact that $P^{**} = P$.

Definition 4.5 (*Dual parameter set*) Let $P = \langle a, t, b, \lambda, \mu \rangle$ be a given parameter set. We denote by P^{*} the parameter set $P^{*} = \langle a, t, b, \mu, \lambda \rangle$. We say that P^{*} is the dual of P.

Our first formula of fractional integration by parts involves the operator K_{P}.

Theorem 4.6 *Let us assume that $k \in L^{q}(\Delta; \mathbb{R})$. Then, the operator $K_{P^{*}}$ defined by*

$$K_{P^{*}}[f](t) = \mu \int\limits_{a}^{t} k(t, \tau) f(\tau) \, d\tau + \lambda \int\limits_{t}^{b} k(\tau, t) f(\tau) \, d\tau \tag{4.5}$$

is a linear bounded operator from $L^{p}(a, b; \mathbb{R})$ to $L^{q}(a, b; \mathbb{R})$. Moreover, the following integration by parts formula holds:

$$\int\limits_{a}^{b} f(t) \cdot K_{P}[g](t) \, dt = \int\limits_{a}^{b} g(t) \cdot K_{P^{*}}[f](t) \, dt, \tag{4.6}$$

for any $f, g \in L^{p}(a, b; \mathbb{R})$.

Proof Using Theorem 4.2, we obtain that $K_{P^{*}}$ is a linear bounded operator from $L^{p}(a, b; \mathbb{R})$ to $L^{q}(a, b; \mathbb{R})$. The second part is easily proved using Fubini's theorem. Indeed, considering only the first term of K_{P}, the following equality holds for any $f, g \in L^{p}(a, b; \mathbb{R})$:

$$\lambda \int\limits_{a}^{b} f(t) \cdot \int\limits_{a}^{t} k(t, \tau) g(\tau) \, d\tau \, dt = \lambda \int\limits_{a}^{b} g(\tau) \cdot \int\limits_{\tau}^{b} k(t, \tau) f(t) \, dt \, d\tau.$$

The proof is completed by using the same strategy on the second part of the definition of K_P.

The next example shows that one cannot relax the hypotheses of Theorem 4.6.

Example 4.7 Let $P = \langle 0, t, 1, 1, -1 \rangle$, $f = g = 1$, and $k(t, \tau) = \frac{t^2 - \tau^2}{(t^2 + \tau^2)^2}$. Direct calculations show that

$$
\int_0^1 K_P[1]\, dt = \int_0^1 \left(\int_0^t \frac{t^2 - \tau^2}{(t^2 + \tau^2)^2}\, d\tau - \int_t^1 \frac{\tau^2 - t^2}{(t^2 + \tau^2)^2}\, d\tau \right) dt
$$

$$
= \int_0^1 \left(\int_0^1 \frac{t^2 - \tau^2}{(t^2 + \tau^2)^2}\, d\tau \right) dt = \int_0^1 \frac{1}{t^2 + 1}\, dt = \frac{\pi}{4}
$$

and

$$
\int_0^1 K_{P*}[1]\, d\tau = \int_0^1 \left(-\int_0^\tau \frac{\tau^2 - t^2}{(t^2 + \tau^2)^2}\, dt + \int_\tau^1 \frac{t^2 - \tau^2}{(t^2 + \tau^2)^2}\, dt \right) d\tau
$$

$$
= -\int_0^1 \left(\int_0^1 \frac{\tau^2 - t^2}{(t^2 + \tau^2)^2}\, dt \right) d\tau = -\int_0^1 \frac{1}{\tau^2 + 1}\, d\tau = -\frac{\pi}{4}.
$$

Therefore, the integration by parts formula (4.6) does not hold. Observe that in this case $\int_0^1 \int_0^1 |k(t, \tau)|^2\, d\tau\, dt = \infty$.

For the classical Riemann–Liouville fractional integrals, Theorem 4.6 gives the following result.

Corollary 4.8 *Let* $\frac{1}{p} < \alpha < 1$. *If* $f, g \in L_p(a, b; \mathbb{R})$, *then*

$$
\int_a^b g(t) \cdot {}_aI_t^\alpha[f](t)\, dt = \int_a^b f(t) \cdot {}_tI_b^\alpha[g](t)\, dt. \tag{4.7}
$$

Proof Let $k^\alpha(t, \tau) = \frac{1}{\Gamma(\alpha)}(t - \tau)^{\alpha - 1}$. Using the same reasoning as in the proof of Corollary 4.3, one has $k^\alpha \in L^q(\Delta; \mathbb{R})$. Therefore, (4.7) follows from (4.6).

Furthermore, Riemann–Liouville integrals of variable order satisfy the following integration by parts formula:

Corollary 4.9 *Let* $\alpha : \Delta \to [\delta, 1]$ *with* $\delta > \frac{1}{p}$ *and let* $f, g \in L_p(a, b; \mathbb{R})$. *Then*

$$\int_a^b g(t) \cdot {_aI_t^{\alpha(\cdot,\cdot)}}[f](t)\, dt = \int_a^b f(t) \cdot {_tI_b^{\alpha(\cdot,\cdot)}}[g](t)\, dt. \tag{4.8}$$

Proof Let $k^\alpha(t, \tau) = \frac{1}{\Gamma(\alpha(t,\tau))}(t - \tau)^{\alpha(t,\tau)-1}$. Similarly as in the proof of Corollary 4.4, for $\alpha : \Delta \to [\delta, 1]$ with $\delta > \frac{1}{p}$ one has $k^\alpha \in L^q(\Delta; \mathbb{R})$. Therefore, (4.8) follows from (4.6).

4.2 Fundamental Problem

For $P = \langle a, t, b, \lambda, \mu \rangle$, let us consider the following functional:

$$\mathcal{I} : \mathcal{A}(y_a, y_b) \longrightarrow \mathbb{R} \tag{4.9}$$
$$y \longmapsto \int_a^b F(y(t), K_P[y](t), \dot{y}(t), B_P[y](t), t)\, dt,$$

where

$$\mathcal{A}(y_a, y_b) := \Big\{ y \in C^1([a, b]; \mathbb{R}) : y(a) = y_a,\ y(b) = y_b,$$
$$K_P[y], B_P[y] \in C([a, b]; \mathbb{R}) \Big\},$$

$\dot{y}$ denotes the classical derivative of y, K_P is the generalized fractional integral operator with kernel belonging to $L^q(\Delta; \mathbb{R})$, $B_P = K_P \circ \frac{d}{dt}$, and F is the Lagrangian function of class C^1:

$$F : \quad \mathbb{R}^4 \times [a, b] \longrightarrow \mathbb{R} \tag{4.10}$$
$$(x_1, x_2, x_3, x_4, t) \longmapsto F(x_1, x_2, x_3, x_4, t).$$

Moreover, we assume that

- $K_{P*}[\tau \mapsto \partial_2 F(y(\tau), K_P[y](\tau), \dot{y}(\tau), B_P[y](\tau), \tau)] \in C([a, b]; \mathbb{R})$,
- $t \mapsto \partial_3 F(y(t), K_P[y](t), \dot{y}(t), B_P[y](t), t) \in C^1([a, b]; \mathbb{R})$,
- $K_{P*}[\tau \mapsto \partial_4 F(y(\tau), K_P[y](\tau), \dot{y}(\tau), B_P[y](\tau), \tau)] \in C^1([a, b]; \mathbb{R})$.

The next result gives a necessary optimality condition of Euler–Lagrange type for the problem of finding a function minimizing functional (4.9).

Theorem 4.10 *Let* $\bar{y} \in \mathcal{A}(y_a, y_b)$ *be a minimizer of functional* (4.9). *Then,* $\bar{y}$ *satisfies the following Euler–Lagrange equation:*

$$\frac{d}{dt}\left[\partial_3 F\left(\star_y\right)(t)\right] + A_{P*}\left[\tau \mapsto \partial_4 F\left(\star_y\right)(\tau)\right](t)$$
$$= \partial_1 F\left(\star_y\right)(t) + K_{P*}\left[\tau \mapsto \partial_2 F\left(\star_y\right)(\tau)\right](t), \qquad (4.11)$$

where $\left(\star_y\right)(t) = (y(t), K_P[y](t), \dot{y}(t), B_P[y](t), t)$, for $t \in (a, b)$.

Proof Because $\bar{y} \in \mathcal{A}(y_a, y_b)$ is a minimizer of (4.9), we have

$$\mathcal{I}(\bar{y}) \le \mathcal{I}(\bar{y} + h\eta),$$

for any $|h| \le \varepsilon$ and every $\eta \in \mathcal{A}(0, 0)$. Let us define the following function:

$$\phi_{\bar{y},\eta} : [-\varepsilon, \varepsilon] \to \mathbb{R}$$
$$h \mapsto \mathcal{I}(\bar{y} + h\eta),$$

where

$$\mathcal{I}(\bar{y} + h\eta) = \int_a^b F(\bar{y}(t) + h\eta(t), K_P[\bar{y} + h\eta](t), \dot{\bar{y}}(t) + h\dot{\eta}(t), B_P[\bar{y} + h\eta](t), t) \, dt.$$

Since $\phi_{\bar{y},\eta}$ is of class C^1 on $[-\varepsilon, \varepsilon]$ and

$$\phi_{\bar{y},\eta}(0) \le \phi_{\bar{y},\eta}(h), \quad |h| \le \varepsilon,$$

we deduce that

$$\phi'_{\bar{y},\eta}(0) = \frac{d}{dh}\mathcal{I}(\bar{y} + h\eta)\bigg|_{h=0} = 0.$$

Hence, by the theorem of differentiation under an integral sign and by the chain rule, we get

$$\int_a^b \left(\partial_1 F(\star_{\bar{y}})(t) \cdot \eta(t) + \partial_2 F(\star_{\bar{y}})(t) \cdot K_P[\eta](t) + \partial_3 F(\star_{\bar{y}})(t) \cdot \dot{\eta}(t)\right.$$
$$\left. + \partial_4 F(\star_{\bar{y}})(t) \cdot B_P[\eta](t)\right) dt = 0.$$

Finally, Theorem 4.6 yields

$$\int_a^b \left(\partial_1 F(\star_{\bar{y}})(t) + K_{P*}\left[\tau \mapsto \partial_2 F(\star_{\bar{y}})(\tau)\right](t)\right) \cdot \eta(t)$$
$$+ \left(\partial_3 F(\star_{\bar{y}})(t) + K_{P*}\left[\tau \mapsto \partial_4 F(\star_{\bar{y}})(\tau)\right](t)\right) \cdot \dot{\eta}(t) \, dt = 0,$$

and applying the classical integration by parts formula and the fundamental lemma of the calculus of variations (see, e.g., (Gelfand and Fomin 2000)) we obtain the validity of (4.11) along $\bar{y}$.

Remark 4.11 From now, in order to simplify the notation, for T, S being fractional operators we will write shortly

$$T \left[\partial_i F(y(\tau), T[y](\tau), \dot{y}(\tau), S[y](\tau), \tau) \right]$$

instead of

$$T \left[\tau \mapsto \partial_i F(y(\tau), T[y](\tau), \dot{y}(\tau), S[y](\tau), \tau) \right], \quad i = 1, \ldots, 5.$$

Example 4.12 Let $P = \langle 0, t, 1, 1, 0 \rangle$. Consider the problem of minimizing the following functional:

$$\mathcal{I}(y) = \int_0^1 (K_P[y](t) + t)^2 \, dt$$

subject to the given boundary conditions

$$y(0) = -1 \quad \text{and} \quad y(1) = -1 - \int_0^1 u(1 - \tau) \, d\tau,$$

where the kernel k of K_P is such that $k(t, \tau) = h(t - \tau)$ with $h \in C^1([0, 1]; \mathbb{R})$, $h(t) \equiv 0$ for $t < 0$, $h(0) = 1$, and we can find $M > 0$ and $\rho > 0$ such that $|h(t)| \leq Me^{\rho t}$, $t \geq 0$. Here the resolvent u is related to the kernel h by $u(t) = \mathcal{L}^{-1} \left[\frac{1}{\tilde{h}(s)} - 1 \right](t)$, $\tilde{h}(s) = \mathcal{L}[h](s)$, where $\mathcal{L}$ and $\mathcal{L}^{-1}$ are the direct and the inverse Laplace operators, respectively. We apply Theorem 4.10 with Lagrangian F given by $F(x_1, x_2, x_3, x_4, t) = (x_2 + t)^2$. Because

$$y(t) = -1 - \int_0^t u(t - \tau) \, d\tau$$

is the solution to the Volterra integral equation of the first kind (see, e.g., Eq. 16, p.114 of Polyanin and Manzhirov (1998))

$$K_P[y](t) + t = 0,$$

it satisfies our generalized Euler–Lagrange equation (4.11), that is,

$$K_{P*} [K_P[y](\tau) + \tau](t) = 0, \ t \in (a, b).$$

In particular, for the kernel $h(t - \tau) = e^{-(t-\tau)}$ and the boundary conditions $y(0) = -1$, $y(1) = -2$, the solution is $y(t) = -1 - t$.

Remark 4.13 (*cf. Theorem 2.2.3 of* van Brunt (2004)) If the functional (4.9) does not depend on K_P and B_P, then Theorem 4.10 reduces to the classical result: if $\bar{y} \in C^2([a, b]; \mathbb{R})$ is a solution to the problem of minimizing the functional

$$\mathcal{I}(y) = \int_a^b F(y(t), \dot{y}(t), t) \, dt, \quad \text{subject to } y(a) = y_a, \quad y(b) = y_b,$$

then $\bar{y}$ satisfies the Euler–Lagrange equation

$$\partial_1 F(y(t), \dot{y}(t), t) - \frac{d}{dt}\partial_2 F(y(t), \dot{y}(t), t) = 0, \quad \text{for all } t \in [a, b].$$

Remark 4.14 In the particular case when functional (4.9) does not depend on the integer derivative of function y, we obtain from Theorem 4.10 the following result: if $\bar{y} \in A(y_a, y_b)$ is a solution to the problem of minimizing the functional

$$\mathcal{I}(y) = \int_a^b F(y(t), K_P[y](t), B_P[y](t), t) \, dt,$$

subject to $y(a) = y_a$ and $y(b) = y_b$, then necessarily

$$\begin{aligned}
A_{P*} &\left[\partial_4 F(\bar{y}(\tau), K_P[\bar{y}](\tau), B_P[\bar{y}](\tau), \tau)\right](t) \\
&= \partial_1 F(\bar{y}(t), K_P[\bar{y}](t), B_P[\bar{y}](t), t) \\
&\quad + K_{P*}\left[\partial_2 F(\bar{y}(\tau), K_P[\bar{y}](\tau), B_P[\bar{y}](\tau), \tau)\right](t), \quad t \in (a, b).
\end{aligned}$$

This extends some of the recent results of Agrawal (2010).

Corollary 4.15 *Let* $0 < \alpha < \frac{1}{q}$ *and let* $\bar{y} \in C^1([a, b]; \mathbb{R})$ *be a solution to the problem of minimizing the functional*

$$\mathcal{I}(y) = \int_a^b F(y(t), {}_aI_t^{1-\alpha}[y](t), \dot{y}(t), {}_a^C D_t^\alpha[y](t), t) \, dt, \tag{4.12}$$

subject to the boundary conditions $y(a) = y_a$ *and* $y(b) = y_b$, *where*

- $F \in C^1(\mathbb{R}^4 \times [a, b]; \mathbb{R})$,
- *functions* $t \mapsto \partial_1 F(y(t), {}_aI_t^{1-\alpha}[y](t), \dot{y}(t), {}_a^C D_t^\alpha[y](t), t)$,
  ${}_tI_b^{1-\alpha}\left[\partial_2 F(y(\tau), {}_aI_\tau^{1-\alpha}[y](\tau), \dot{y}(\tau), {}_a^C D_\tau^\alpha[y](\tau), \tau)\right]$ *are continuous on* $[a, b]$,

- *functions* $t \mapsto \partial_3 F(y(t), {}_aI_t^{1-\alpha}[y](t), \dot{y}(t), {}_a^C D_t^\alpha[y](t), t)$,
  $_tI_b^{1-\alpha}\left[\partial_4 F(y(\tau), {}_aI_\tau^{1-\alpha}[y](\tau), \dot{y}(\tau), {}_a^C D_\tau^\alpha[y](\tau), \tau)\right]$ *are continuously differentiable on* $[a, b]$.

Then, the following holds:

$$\frac{d}{dt}\left(\partial_3 F(\bar{y}(t), {}_aI_t^{1-\alpha}[\bar{y}](t), \dot{\bar{y}}(t), {}_a^C D_t^\alpha[\bar{y}](t), t)\right)$$

$$- {}_tD_b^\alpha\left[\partial_4 F(\bar{y}(\tau), {}_aI_\tau^{1-\alpha}[\bar{y}](\tau), \dot{\bar{y}}(\tau), {}_a^C D_\tau^\alpha[\bar{y}](\tau), \tau)\right](t)$$

$$= \partial_1 F(\bar{y}(t), {}_aI_t^{1-\alpha}[\bar{y}](t), \dot{\bar{y}}(t), {}_a^C D_t^\alpha[\bar{y}](t), t)$$

$$+ {}_tI_b^\alpha\left[\partial_2 F(\bar{y}(\tau), {}_aI_\tau^{1-\alpha}[\bar{y}](\tau), \dot{\bar{y}}(\tau), {}_a^C D_\tau^\alpha[\bar{y}](\tau), \tau)\right](t), \quad t \in (a, b).$$

$$(4.13)$$

Proof The intended Euler–Lagrange equation follows from (4.11) by choosing $P = \langle a, t, b, 1, 0 \rangle$ and the kernel $k^\alpha(t, \tau) = \frac{1}{\Gamma(1-\alpha)}(t - \tau)^{-\alpha}$. Note that for $0 < \alpha < \frac{1}{q}$, we have $k^\alpha \in L^q(\Delta; \mathbb{R})$.

In Example 4.16 we shall make use of the Mittag–Leffler function of one parameter. Let $\alpha > 0$. We recall that the Mittag–Leffler function is defined by

$$E_\alpha(z) = \sum_{k=0}^\infty \frac{z^k}{\Gamma(\alpha k + 1)},$$

where $z \in \mathbb{C}$. This function appears naturally in the solutions of fractional differential equations, as a generalization of the exponential function (Camargo et al. 2009). Indeed, while a linear ordinary differential equation with constant coefficients presents an exponential function in its solution, in the fractional case the Mittag–Leffler functions emerge (Kilbas et al. 2006).

Example 4.16 Let $0 < \alpha < \frac{1}{q}$. Consider the problem of minimizing the functional

$$\mathcal{I}(y) = \int_0^1 \sqrt{1 + (\dot{y}(t) + {}_a^C D_t^\alpha[y](t) - 1)^2} \, dt \qquad (4.14)$$

subject to the following boundary conditions:

$$y(0) = 0 \quad \text{and} \quad y(1) = \int_0^1 E_{1-\alpha}\left[-(1 - \tau)^{1-\alpha}\right] d\tau. \qquad (4.15)$$

Function F of Corollary 4.15 is given by

$$F(x_1, x_2, x_3, x_4, t) = \sqrt{1 + (x_3 + x_4 - 1)^2}.$$

One can easily check that (see Kilbas et al. (2006) p. 324)

$$y(t) = \int_0^t E_{1-\alpha}\left[-(t-\tau)^{1-\alpha}\right] d\tau \qquad (4.16)$$

satisfies $\dot{y}(t) + {}_a^C D_t^\alpha[y](t) \equiv 1$. Moreover, it satisfies

$$\frac{d}{dt}\left(\frac{\dot{y}(t) + {}_a^C D_t^\alpha[y](t) - 1}{\sqrt{1 + (\dot{y}(t) + {}_a^C D_t^\alpha[y](t) - 1)^2}}\right)$$

$$- {}_tD_b^\alpha\left[\frac{\dot{y}(\tau) + {}_a^C D_\tau^\alpha[y](\tau) - 1}{\sqrt{1 + (\dot{y}(\tau) + {}_a^C D_\tau^\alpha[y](\tau) - 1)^2}}\right](t) = 0,$$

for all $t \in (a, b)$. We conclude that (4.16) is a candidate function for giving a minimum to problem (4.14) and (4.15).

Corollary 4.17 (cf. Agrawal 2007) *Let* $0 < \alpha < \frac{1}{q}$, $\mathcal{I}$ *be the functional*

$$\mathcal{I}(y) = \int_a^b F\left(y(t), \dot{y}(t), \lambda {}_a^C D_t^\alpha[y](t) + \mu {}_t^C D_b^\alpha[y](t), t\right) dt, \qquad (4.17)$$

where λ and μ are real numbers, and $\bar{y} \in C^1([a, b]; \mathbb{R})$ be a minimizer of $\mathcal{I}$ among all functions satisfying boundary conditions $y(a) = y_a$, $y(b) = y_b$. Moreover, we assume that

- $F \in C^1(\mathbb{R}^3 \times [a, b]; \mathbb{R})$,
- *functions* $t \mapsto \partial_2 F\left(y(t), \dot{y}(t), \lambda {}_a^C D_t^\alpha[y](t) + \mu {}_t^C D_b^\alpha[y](t), t\right)$,
  ${}_aI_t^{1-\alpha}\left[\partial_3 F\left(y(\tau), \dot{y}(\tau), \lambda {}_a^C D_\tau^\alpha[y](\tau) + \mu {}_\tau^C D_b^\alpha[y](\tau), \tau\right)\right]$,
  and ${}_tI_b^{1-\alpha}\left[\partial_3 F\left(y(\tau), \dot{y}(\tau), \lambda {}_a^C D_\tau^\alpha[y](\tau) + \mu {}_\tau^C D_b^\alpha[y](\tau), \tau\right)\right]$ *are continuously differentiable on $[a, b]$.*

Then, $\bar{y}$ satisfies the Euler–Lagrange equation

$$\lambda {}_tD_b^\alpha\left[\partial_3 F\left(y(\tau), \dot{y}(\tau), \lambda {}_a^C D_\tau^\alpha[y](\tau) + \mu {}_\tau^C D_b^\alpha[y](\tau), \tau\right)\right](t)$$

$$+ \mu {}_aD_t^\alpha\left[\partial_3 F\left(y(\tau), \dot{y}(\tau), \lambda {}_a^C D_\tau^\alpha[y](\tau) + \mu {}_\tau^C D_b^\alpha[y](\tau), \tau\right)\right](t)$$

$$+ \partial_1 F\left(y(t), \dot{y}(t), \lambda {}_a^C D_t^\alpha[y](t) + \mu {}_t^C D_b^\alpha[y](t), t\right)$$

$$- \frac{d}{dt}\left(\partial_2 F\left(y(t), \dot{y}(t), \lambda {}_a^C D_t^\alpha[y](t) + \mu {}_t^C D_b^\alpha[y](t), t\right)\right) = 0 \qquad (4.18)$$

for all $t \in (a, b)$.

Proof Choose $P = \langle a, t, b, \lambda, -\mu \rangle$ and $k^\alpha(t - \tau) = \frac{1}{\Gamma(1-\alpha)}(t - \tau)^{-\alpha}$. Then, for $0 < \alpha < \frac{1}{q}$ kernel k^α is in $L^q(\Delta; \mathbb{R})$, the operator B_P reduces to the sum of the left and right Caputo fractional derivatives and (4.18) follows from (4.11).

Corollary 4.18 *Let us consider the problem of minimizing a functional*

$$\mathcal{I}(y) = \int_a^b F\left(y(t), {}_aI_t^{1-\alpha(\cdot,\cdot)}[y](t), \dot{y}(t), {}_a^C D_t^{\alpha(\cdot,\cdot)}[y](t), t\right) dt \qquad (4.19)$$

subject to boundary conditions

$$y(a) = y_a, \quad y(b) = y_b, \qquad (4.20)$$

where $\dot{y}, {}_aI_t^{1-\alpha(\cdot,\cdot)}[y], {}_a^C D_t^{\alpha(\cdot,\cdot)}[y] \in C([a, b]; \mathbb{R})$ *and* $\alpha : \Delta \to [0, 1 - \delta]$ *with* $\delta > \frac{1}{p}$. *Moreover, we assume that*

- $F \in C^1(\mathbb{R}^4 \times [a, b], \mathbb{R})$,
- *function* ${}_tI_b^{1-\alpha(\cdot,\cdot)}\left[\partial_2 F\left(y(\tau), {}_aI_\tau^{1-\alpha(\cdot,\cdot)}[y](\tau), \dot{y}(\tau), {}_a^C D_\tau^{\alpha(\cdot,\cdot)}[y](\tau), \tau\right)\right]$ *is continuous on* $[a, b]$,
- *functions* $t \mapsto \partial_3 F\left(y(t), {}_aI_t^{1-\alpha(\cdot,\cdot)}[y](t), \dot{y}(t), {}_a^C D_t^{\alpha(\cdot,\cdot)}[y](t), t\right)$
  and ${}_tI_b^{1-\alpha(\cdot,\cdot)}\left[\partial_4 F\left(y(\tau), {}_aI_\tau^{1-\alpha(\cdot,\cdot)}[y](\tau), \dot{y}(\tau), {}_a^C D_\tau^{\alpha(\cdot,\cdot)}[y](\tau), \tau\right)\right]$ *are continuously differentiable on* $[a, b]$.

Then, if $\bar{y} \in C^1([a, b]; \mathbb{R})$ *is a solution to problem* (4.19) *and* (4.20), *it necessarily satisfies the Euler–Lagrange equation*

$$\partial_1 F\left(y(t), {}_aI_t^{1-\alpha(\cdot,\cdot)}[y](t), \dot{y}(t), {}_a^C D_t^{\alpha(\cdot,\cdot)}[y](t), t\right)$$
$$- \frac{d}{dt}\partial_3 F\left(y(t), {}_aI_t^{1-\alpha(\cdot,\cdot)}[y](t), \dot{y}(t), {}_a^C D_t^{\alpha(\cdot,\cdot)}[y](t), t\right)$$
$$+ {}_tI_b^{1-\alpha(\cdot,\cdot)}\left[\partial_2 F\left(y(\tau), {}_aI_\tau^{1-\alpha(\cdot,\cdot)}[y](\tau), \dot{y}(\tau), {}_a^C D_\tau^{\alpha(\cdot,\cdot)}[y](\tau), \tau\right)\right](t)$$
$$+ {}_tD_b^{\alpha(\cdot,\cdot)}\left[\partial_4 F\left(y(\tau), {}_aI_\tau^{1-\alpha(\cdot,\cdot)}[y](\tau), \dot{y}(\tau), {}_a^C D_\tau^{\alpha(\cdot,\cdot)}[y](\tau), \tau\right)\right](t) = 0$$
$$\qquad (4.21)$$

for all $t \in (a, b)$.

Proof For $\alpha : \Delta \to [0, 1 - \delta]$ with $\delta > \frac{1}{p}$ we have that $k^\alpha(t, \tau) = \frac{1}{\Gamma(1-\alpha(t,\tau))}$ $(t - \tau)^{-\alpha(t,\tau)}$ is in $L^q(\Delta; \mathbb{R})$. Therefore, from (4.11) follows (4.21).

In the next example $\alpha(t)$ is a function defined on the interval $[a, b]$ and taking values in the set $[0, 1-\delta]$, where $\delta > \frac{1}{p}$. As before, we assume that $\alpha \in C^1([a, b]; \mathbb{R})$.

Example 4.19 Consider the following problem:

$$J(y) = \int\limits_a^b \left({}_a^C D_t^{\alpha(\cdot)}[y](t) \right)^2 + \left({}_a I_t^{1-\alpha(\cdot)}[y](t) - \frac{\xi(t-\tau)^{1-\alpha(t)}}{\Gamma(2-\alpha(t))} \right)^2 dt \longrightarrow \min,$$

$$y(a) = \xi, \quad y(b) = \xi,$$

for a given real ξ. Because $J(y) \geq 0$ for any function y and $J(\bar{y}) = 0$ for the admissible function $\bar{y} = \xi$ (use relation (2.11) for $\gamma = 0$, the linearity of operator ${}_a I_t^{1-\alpha(\cdot)}$, and the definition of left Caputo derivative of a variable fractional order), we conclude that $\bar{y}$ is the global minimizer to the problem. It is straightforward to check that $\bar{y}$ satisfies our variable order fractional Euler–Lagrange equation (4.21).

Next result gives a sufficient condition assuring that solution of (4.11) is indeed minimizer of (4.9).

Theorem 4.20 *Let $\bar{y} \in \mathcal{A}(y_a, y_b)$ satisfies (4.11) and $(x_1, x_2, x_3, x_4) \longmapsto F(x_1, x_2, x_3, x_4, t)$ be convex for every $t \in [a, b]$. Then $\bar{y}$ is a minimizer of (4.9).*

Proof Let us assume that $\bar{y} \in \mathcal{A}(y_a, y_b)$ satisfies Eq. (4.11) and that $(x_1, x_2, x_3, x_4) \longmapsto F(x_1, x_2, x_3, x_4, t)$ is convex for every $t \in [a, b]$. Then, for every $y \in \mathcal{A}(y_a, y_b)$ we have

$$\mathcal{I}(y) \geq \mathcal{I}(\bar{y}) + \int\limits_a^b (\partial_1 F \cdot (y - \bar{y}) + \partial_2 F \cdot (K_P[y] - K_P[\bar{y}])$$

$$+ \partial_3 F \cdot (\dot{y} - \dot{\bar{y}}) + \partial_4 F \cdot (B_P[y] - B_P[\bar{y}])) \, dt,$$

where $\partial_i F$ are taken in $(\bar{y}, K_P[\bar{y}], \dot{\bar{y}}, B_P[\bar{y}], t), i = 1, 2, 3, 4$. Having in mind that $y(a) - \bar{y}(a) = y(b) - \bar{y}(b) = 0$, and using the classical integration by parts formula as well as Theorem 4.6, one has

$$\mathcal{I}(y) \geq \mathcal{I}(\bar{y}) + \int\limits_a^b \left(\partial_1 F + K_{P*}[\partial_2 F] - \frac{d}{dt}(\partial_3 F) - A_{P*}[\partial_4 F] \right) (y - \bar{y}) \, dt,$$

where, as before, $\partial_i F$ are taken in $(\bar{y}, K_P[\bar{y}], \dot{\bar{y}}, B_P[\bar{y}], t), i = 1, 2, 3, 4$. Finally, by Euler–Lagrange equation (4.11), we have $\mathcal{I}(y) \geq \mathcal{I}(\bar{y})$.

4.3 Free Initial Boundary

Let us define the set

$$\mathcal{A}(y_b) := \Big\{ y \in C^1([a, b]; \mathbb{R}) : y(a) \text{ is free}, \quad y(b) = y_b,$$

$$K_P[y], B_P[y] \in C([a, b]; \mathbb{R}) \Big\},$$

and let $\bar{y}$ be a minimizer of functional (4.9) on $\mathcal{A}(y_b)$, i.e., now

$$\mathcal{I} : \mathcal{A}(y_b) \longrightarrow \mathbb{R} \tag{4.22}$$

$$y \longmapsto \int_a^b F(y(t), K_P[y](t), \dot{y}(t), B_P[y](t), t) \, dt.$$

Because

$$\mathcal{I}(\bar{y}) \leq \mathcal{I}(\bar{y} + h\eta),$$

for any $|h| \leq \varepsilon$ and every $\eta \in \mathcal{A}(0)$, we obtain as in the proof of Theorem 4.10 that

$$\int_a^b \big(\partial_1 F(\star_{\bar{y}})(t) + K_{P*}\big[\partial_2 F(\star_{\bar{y}})(\tau)\big](t)\big) \cdot \eta(t)$$

$$+ \big(\partial_3 F(\star_{\bar{y}})(t) + K_{P*}\big[\partial_4 F(\star_{\bar{y}})(\tau)\big](t)\big) \cdot \dot{\eta}(t) \, dt = 0, \quad \forall \eta \in \mathcal{A}(0),$$

where $(\star_{\bar{y}})(t) = (\bar{y}(t), K_P[\bar{y}](t), \dot{\bar{y}}(t), B_P[\bar{y}](t), t)$. Moreover, having in mind that $\eta(b) = 0$ and using the classical integration by parts formula, we find that

$$\int_a^b \Bigg(\partial_1 F(\star_{\bar{y}})(t) + K_{P*}\big[\partial_2 F(\star_{\bar{y}})(\tau)\big](t)$$

$$- \frac{d}{dt}\big(\partial_3 F(\star_{\bar{y}})(t) + K_{P*}\big[\partial_4 F(\star_{\bar{y}})(\tau)\big](t)\big)\Bigg) \cdot \eta(t) \, dt$$

$$+ \partial_3 F(\star_{\bar{y}})(t) \cdot \eta(t)\big|_a + K_{P*}\big[\partial_4 F(\star_{\bar{y}})(\tau)\big](t) \cdot \eta(t)\big|_a = 0$$

for all $\eta \in \mathcal{A}(0)$. Now, using the fundamental lemma of the calculus of variations (see, e.g., (Gelfand and Fomin 2000)) and the fact that $\eta(a)$ is arbitrary, we obtain

$$\begin{cases} \frac{d}{dt}\big[\partial_3 F(\star_{\bar{y}})(t)\big] + A_{P*}\big[\partial_4 F(\star_{\bar{y}})(\tau)\big](t) = \partial_1 F(\star_{\bar{y}})(t) + K_{P*}\big[\partial_2 F(\star_{\bar{y}})(\tau)\big](t), \\ \partial_3 F(\star_{\bar{y}})(t)\big|_a + K_{P*}\big[\partial_4 F(\star_{\bar{y}})(\tau)\big](t)\big|_a = 0. \end{cases}$$

We have just proved the following result.

Theorem 4.21 *If $\bar{y} \in \mathcal{A}(y_b)$ is a solution to the problem of minimizing functional* (4.22) *on the set $\mathcal{A}(y_b)$, then $\bar{y}$ satisfies the Euler–Lagrange equation* (4.11). *Moreover, the extra boundary condition*

$$\partial_3 F(\star_{\bar{y}})(t)\big|_a + K_{P^*}\left[\partial_4 F(\star_{\bar{y}})(\tau)\right](t)\big|_a = 0 \tag{4.23}$$

holds with $(\star_{\bar{y}})(t) = (\bar{y}(t), K_P[\bar{y}](t), \dot{\bar{y}}(t), B_P[\bar{y}](t), t)$.

Similarly as it is in the theory of the classical calculus of variations, we will call (4.23) the generalized fractional natural boundary condition.

Corollary 4.22 (cf. Agrawal 2006) *Let* $0 < \alpha < \frac{1}{q}$ *and $\mathcal{I}$ be the functional given by*

$$\mathcal{I}(y) = \int\limits_a^b F\left(y(t), {}^C_a D_t^\alpha[y](t), t\right) dt,$$

where $F \in C^1(\mathbb{R}^2 \times [a, b]; \mathbb{R})$, *and* ${}_a I_t^{1-\alpha}\left[\partial_2 F\left(y(\tau), {}^C_a D_\tau^\alpha[y](\tau), \tau\right)\right]$ *is continuously differentiable on* $[a, b]$. *If* $\bar{y} \in C^1([a, b]; \mathbb{R})$ *is a minimizer of $\mathcal{I}$ among all functions satisfying the boundary condition* $y(b) = y_b$, *then $\bar{y}$ satisfies the Euler–Lagrange equation*

$$\partial_1 F\left(y(t), {}^C_a D_t^\alpha[y](t), t\right) + {}_t D_b^\alpha\left[\partial_2 F\left(y(\tau), {}^C_a D_\tau^\alpha[y](\tau), \tau\right)\right](t) = 0$$

for all $t \in (a, b)$, *and the fractional natural boundary condition*

$${}_t I_b^{1-\alpha}\left[\partial_2 F\left(\bar{y}(\tau), {}^C_a D_\tau^\alpha[\bar{y}](\tau), \tau\right)\right](t)\big|_a = 0$$

holds.

Proof Corollary 4.22 follows from Theorem 4.21 with $P = \langle a, t, b, 1, 0 \rangle$ and $k^\alpha(t, \tau) = \frac{1}{\Gamma(1-\alpha)}(t - \tau)^{-\alpha}$.

Corollary 4.23 *Suppose that* $\alpha : \Delta \to [0, 1 - \delta]$, $\delta > \frac{1}{p}$ *and $\mathcal{I}$ is the functional given by* (4.19). *If* $\bar{y} \in C^1([a, b]; \mathbb{R})$ *is a minimizer to $\mathcal{I}$ satisfying boundary condition* $y(b) = y_b$ *and being such that* ${}_a I_t^{1-\alpha(\cdot, \cdot)}[y], {}^C_a D_t^{\alpha(\cdot, \cdot)}[y] \in C([a, b]; \mathbb{R})$, *then*

$$\partial_1 F\left(\bar{y}(t), {}_a I_t^{1-\alpha(\cdot, \cdot)}[\bar{y}](t), \dot{\bar{y}}, {}^C_a D_t^{\alpha(\cdot, \cdot)}[\bar{y}](t), t\right)$$

$$- \frac{d}{dt} \partial_3 F\left(\bar{y}(t), {}_a I_t^{1-\alpha(\cdot, \cdot)}[\bar{y}](t), \dot{\bar{y}}(t), {}^C_a D_t^{\alpha(\cdot, \cdot)}[\bar{y}](t), t\right)$$

$$+ {}_t I_b^{1-\alpha(\cdot, \cdot)}\left[\partial_2 F\left(\bar{y}(\tau), {}_a I_\tau^{1-\alpha(\cdot, \cdot)}[\bar{y}](\tau), \dot{y}(\tau), {}^C_a D_\tau^{\alpha(\cdot, \cdot)}[\bar{y}](\tau), \tau\right)\right](t)$$

$$+ {}_t D_b^{\alpha(\cdot, \cdot)}\left[\partial_4 F\left(\bar{y}(\tau), {}_a I_\tau^{1-\alpha(\cdot, \cdot)}[\bar{y}](\tau), \dot{\bar{y}}(\tau), {}^C_a D_\tau^{\alpha(\cdot, \cdot)}[\bar{y}](\tau), \tau\right)\right](t) = 0$$

for all $t \in (a, b)$, and the natural boundary condition

$$\partial_3 F \left(\bar{y}(t), {}_aI_t^{1-\alpha(\cdot,\cdot)}[\bar{y}](t), \dot{\bar{y}}(t), {}_a^C D_t^{\alpha(\cdot,\cdot)}[\bar{y}](t), t \right) \Big|_a$$

$$+ {}_tI_b^{1-\alpha(\cdot,\cdot)} \left[\partial_4 F \left(\bar{y}(\tau), {}_aI_\tau^{1-\alpha(\cdot,\cdot)}[\bar{y}](\tau), \dot{\bar{y}}(\tau), {}_a^C D_\tau^{\alpha(\cdot,\cdot)}[\bar{y}](\tau), \tau \right) \right] (t) \Big|_a = 0$$

holds.

Proof Corollary 4.23 is an easy consequence of Theorem 4.21.

Remark 4.24 Observe that if the functional (4.22) is independent of the operator K_P, that is, we have the problem

$$\int_a^b F\left(y(t), \dot{y}(t), B_P[y](t), t\right) \, \mathrm{d}t \longrightarrow \min, \quad y(b) = y_b$$

($y(a)$ free), then the optimality conditions (4.11) and (4.23) reduce, respectively, to

$$\partial_1 F\left(\bar{y}(t), \dot{\bar{y}}(t), B_P[\bar{y}](t), t\right) - \frac{d}{\mathrm{d}t} \partial_2 F\left(\bar{y}(t), \dot{\bar{y}}(t), B_P[\bar{y}](t), t\right)$$

$$- A_{P*}\left[\partial_3 F\left(\bar{y}(\tau), \dot{\bar{y}}(\tau), B_P[\bar{y}](\tau), \tau\right)\right](t) = 0$$

for all $t \in (a, b)$, and

$$\partial_2 F\left(\bar{y}(t), \dot{\bar{y}}(t), B_P[\bar{y}](t), t\right)\Big|_a + K_{P*}\left[\partial_3 F\left(\bar{y}(\tau), \dot{\bar{y}}(\tau), B_P[\bar{y}](\tau), \tau\right)\right](t)\Big|_a = 0.$$

4.4 Isoperimetric Problem

One of the earliest problems in geometry is the isoperimetric problem, already considered by the ancient Greeks. It consists to find, among all closed curves of a given length, the one which encloses the maximum area. The general problem for which one integral is to be given a fixed value, while another is to be made a maximum or a minimum, is nowadays part of the calculus of variations (Giaquinta and Hildebrandt 2004; van Brunt 2004). Such *isoperimetric problems* have found a broad class of important applications throughout the centuries, with numerous useful implications in astronomy, geometry, algebra, analysis, and engineering (Blasjo 2005; Curtis 2004). For recent advancements on the study of isoperimetric problems see Almeida and Torres (2009a, b), Ferreira and Torres (2010) and references therein. Here we consider isoperimetric problems with generalized fractional operators. Similarly to Sect. 4.2 and 4.3, we deal with integrands involving both the generalized Caputo fractional derivative and the generalized fractional integral, as well as the classical derivative.

Let $P = \langle a, t, b, \lambda, \mu \rangle$. We define the following functional:

$$\mathcal{J} : \mathcal{A}(y_a, y_b) \longrightarrow \mathbb{R} \tag{4.24}$$
$$y \longmapsto \int_a^b G(y(t), K_P[y](t), \dot{y}(t), B_P[y](t), t)\, dt,$$

where by $\dot{y}$ we understand the classical derivative of y, K_P is the generalized fractional integral operator with kernel belonging to $L^q(\Delta; \mathbb{R})$, $B_P = K_P \circ \frac{d}{dt}$ and G is a Lagrangian of class C^1:

$$G : \quad \mathbb{R}^4 \times [a, b] \longrightarrow \mathbb{R}$$
$$(x_1, x_2, x_3, x_4, t) \longmapsto G(x_1, x_2, x_3, x_4, t).$$

Moreover, we assume that

- $K_{P*}\left[\partial_2 G(y(\tau), K_P[y](\tau), \dot{y}(\tau), B_P[y](\tau), \tau)\right] \in C([a, b]; \mathbb{R})$,
- $t \mapsto \partial_3 G(y(t), K_P[y](t), \dot{y}(t), B_P[y](t), t) \in C^1([a, b]; \mathbb{R})$,
- $K_{P*}\left[\partial_4 G(y(\tau), K_P[y](\tau), \dot{y}(\tau), B_P[y](\tau), \tau)\right] \in C^1([a, b]; \mathbb{R})$.

The first problem in this section is to find a minimizer of functional (4.9) subject to the isoperimetric constraint $\mathcal{J}(y) = \xi$. In the next theorem, we provide a necessary optimality condition for this type of problem.

Theorem 4.25 *Suppose that $\bar{y}$ is a minimizer of functional $\mathcal{I}$ in the class*

$$\mathcal{A}_\xi(y_a, y_b) := \{y \in \mathcal{A}(y_a, y_b) : \mathcal{J}(y) = \xi\}.$$

Then there exists a real constant λ_0, such that, for $H = F - \lambda_0 G$, equality

$$\frac{d}{dt}\left[\partial_3 H(\star_{\bar{y}})(t)\right] + A_{P*}\left[\partial_4 H(\star_{\bar{y}})(\tau)\right](t)$$
$$= \partial_1 H(\star_{\bar{y}})(t) + K_{P*}\left[\partial_2 H(\star_{\bar{y}})(\tau)\right](t), \tag{4.25}$$

$t \in (a, b)$, holds, provided that

$$\frac{d}{dt}\left[\partial_3 G(\star_{\bar{y}})(t)\right] + A_{P*}\left[\partial_4 G(\star_{\bar{y}})(\tau)\right](t)$$
$$\neq \partial_1 G(\star_{\bar{y}})(t) + K_{P*}\left[\partial_2 G(\star_{\bar{y}})(\tau)\right](t), \tag{4.26}$$

$t \in (a, b)$, where $(\star_{\bar{y}})(t) = (\bar{y}(t), K_P[\bar{y}](t), \dot{\bar{y}}(t), B_P[\bar{y}](t), t)$.

Proof By hypothesis (4.26) and the fundamental lemma of the calculus of variations (see, e.g., (Gelfand and Fomin 2000)) we can choose $\eta_2 \in \mathcal{A}(0, 0)$ so that

$$\int\limits_a^b \left(\partial_1 G(\star_{\bar y})(t) + K_{P*}\left[\partial_2 G(\star_{\bar y})(\tau) \right](t) \right) \cdot \eta_2(t)$$

$$+ \left(\partial_3 G(\star_{\bar y})(t) + K_{P*}\left[\partial_4 G(\star_{\bar y})(\tau) \right](t) \right) \cdot \dot\eta_2(t) \, dt = 1.$$

With this function η_2 and an arbitrary $\eta_1 \in \mathcal{A}(0,0)$, let us define functions

$$\phi : [-\varepsilon_1, \varepsilon_1] \times [-\varepsilon_2, \varepsilon_2] \longrightarrow \mathbb{R}$$
$$(h_1, h_2) \longmapsto \mathcal{I}(\bar y + h_1 \eta_1 + h_2 \eta_2)$$

and

$$\psi : [-\varepsilon_1, \varepsilon_1] \times [-\varepsilon_2, \varepsilon_2] \longrightarrow \mathbb{R}$$
$$(h_1, h_2) \longmapsto \mathcal{J}(\bar y + h_1 \eta_1 + h_2 \eta_2) - \xi.$$

Observe that $\psi(0,0) = 0$ and

$$\left.\frac{\partial \psi}{\partial h_2}\right|_{(0,0)} = \int\limits_a^b \left(\partial_1 G(\star_{\bar y})(t) + K_{P*}\left[\partial_2 G(\star_{\bar y})(\tau) \right](t) \right) \cdot \eta_2(t)$$

$$+ \left(\partial_3 G(\star_{\bar y})(t) + K_{P*}\left[\partial_4 G(\star_{\bar y})(\tau) \right](t) \right) \cdot \dot\eta_2(t) \, dt = 1.$$

According to the implicit function theorem we can find $\epsilon_0 > 0$ and a function $s \in C^1([-\varepsilon_0, \varepsilon_0]; \mathbb{R})$ with $s(0) = 0$ such that

$$\psi(h_1, s(h_1)) = 0, \quad \forall h_1 \in [-\varepsilon_0, \varepsilon_0]$$

which implies that $\bar y + h_1 \eta_1 + s(h_1)\eta_2 \in \mathcal{A}_\xi(y_a, y_b)$. We also have

$$\frac{\partial \psi}{\partial h_1} + \frac{\partial \psi}{\partial h_2} \cdot s'(h_1) = 0, \quad \forall h_1 \in [-\varepsilon_0, \varepsilon_0]$$

and hence

$$s'(0) = -\left.\frac{\partial \psi}{\partial h_1}\right|_{(0,0)}.$$

Because $\bar y \in \mathcal{A}(y_a, y_b)$ is a minimizer of $\mathcal{I}$ we have

$$\phi(0,0) \leq \phi(h_1, s(h_1)), \quad \forall h_1 \in [-\varepsilon_0, \varepsilon_0]$$

and then

$$\left.\frac{\partial \phi}{\partial h_1}\right|_{(0,0)} + \left.\frac{\partial \phi}{\partial h_2}\right|_{(0,0)} \cdot s'(0) = 0.$$

Letting $\lambda_0 = \left.\frac{\partial \phi}{\partial h_2}\right|_{(0,0)}$ be the Lagrange multiplier we find

$$\frac{\partial \phi}{\partial h_1}\bigg|_{(0,0)} - \lambda_0 \frac{\partial \psi}{\partial h_1}\bigg|_{(0,0)} = 0$$

or, in other words,

$$\int_a^b \{[(\partial_1 F(\star_{\bar{y}})(t) + K_{P*}[\partial_2 F(\star_{\bar{y}})(\tau)](t)) \cdot \eta_1(t) + (\partial_3 F(\star_{\bar{y}})(t)$$

$$+ K_{P*}[\partial_4 F(\star_{\bar{y}})(\tau)](t)) \cdot \dot{\eta}_1(t)]$$

$$- \lambda_0 [(\partial_1 G(\star_{\bar{y}})(t) + K_{P*}[\partial_2 G(\star_{\bar{y}})(\tau)](t)) \cdot \eta_1(t)$$

$$+ (\partial_3 G(\star_{\bar{y}})(t) + K_{P*}[\partial_4 G(\star_{\bar{y}})(\tau)](t)) \cdot \dot{\eta}_1(t)]\} \, dt = 0.$$

Finally, applying one more time the fundamental lemma of the calculus of variations we obtain (4.25).

Example 4.26 Let $P = \langle 0, t, 1, 1, 0 \rangle$. Consider the problem

$$\mathcal{I}(y) = \int_0^1 (K_P[y](t) + t)^2 \, dt \longrightarrow \min,$$

$$\mathcal{J}(y) = \int_0^1 t K_P[y](t) \, dt = \xi,$$

$$y(0) = \xi - 1, y(1) = (\xi - 1)\left(1 + \int_0^1 u(1 - \tau) \, d\tau\right),$$

where the kernel k is such that $k(t, \tau) = h(t - \tau)$ with $h \in C^1([0, 1]; \mathbb{R})$, $h(t) \equiv 0$ for $t < 0$, $h(0) = 1$, and we can find $M > 0$ and $\rho > 0$ such that $|h(t)| \le M e^{\rho t}$, $t \ge 0$. Moreover, we assume that $K_{P*}[id](t) \ne 0$ (notation id means identity transformation, i.e., $id(t) = t$). Here the resolvent u is related to the kernel h in the same way as in Example 4.12. Since $K_{P*}[id](t) \ne 0$, there is no solution to the Euler–Lagrange equation for functional $\mathcal{J}$. The augmented Lagrangian H of Theorem 4.25 is given by $H(x_1, x_2, t) = (x_2 + t)^2 - \lambda_0 t x_2$. Function

$$y(t) = (\xi - 1)\left(1 + \int_0^t u(t - \tau) \, d\tau\right)$$

is the solution to the Volterra integral equation of the first kind $K_P[y](t) = (\xi - 1)t$ (see, e.g., Eq. 16, p. 114 of Polyanin and Manzhirov (1998)) and for $\lambda_0 = 2\xi$ satisfies our optimality condition (4.25):

$$K_{P^*} \left[2 \left(K_P[y](\tau) + \tau \right) - 2\xi\tau \right] (t) = 0. \tag{4.27}$$

The solution of (4.27) subject to the given boundary conditions depends on the particular choice for the kernel. For example, let $h^\alpha(t - \tau) = e^{\alpha(t-\tau)}$. Then the solution of (4.27) subject to the boundary conditions $y(0) = \xi - 1$ and $y(1) = (\xi - 1)(1 - \alpha)$ is $y(t) = (\xi - 1)(1 - \alpha t)$ (cf. p. 15 of Polyanin and Manzhirov (1998)). If $h^\alpha(t - \tau) = \cos(\alpha(t - \tau))$, then the boundary conditions are $y(0) = \xi - 1$ and $y(1) = (\xi - 1) \left(1 + \alpha^2/2 \right)$, and the extremal is $y(t) = (\xi - 1) \left(1 + \alpha^2 t^2/2 \right)$ (cf. p. 46 of Polyanin and Manzhirov (1998)).

Borrowing different kernels from the book (Polyanin and Manzhirov 1998), many other examples of dynamic optimization problems can be explicitly solved by application of the results of this section.

As particular cases of our problem (4.9) and (4.24), one obtains previously studied fractional isoperimetric problems with Caputo derivatives.

Corollary 4.27 (cf. (Odzijewicz et al. 2012b)) *Let $\bar{y} \in C^1([a, b]; \mathbb{R})$ be a minimizer to the functional*

$$\mathcal{I}(y) = \int_a^b F\left(y(t), \dot{y}(t), {}_a^C D_t^\alpha[y](t), t \right) dt$$

subject to an isoperimetric constraint of the form

$$\mathcal{J}(y) = \int_a^b G\left(y(t), \dot{y}(t), {}_a^C D_t^\alpha[y](t), t \right) dt = \xi,$$

and boundary conditions

$$y(a) = y_a, \quad y(b) = y_b,$$

where $0 < \alpha < \frac{1}{q}$, and functions F, G are such that

- *$F, G \in C^1(\mathbb{R}^3 \times [a, b]; \mathbb{R})$,*
- *$t \mapsto \partial_2 F\left(y(t), \dot{y}(t), {}_a^C D_t^\alpha[y](t), t \right)$, $t \mapsto \partial_2 G\left(y(t), \dot{y}(t), {}_a^C D_t^\alpha[y](t), t \right)$, ${}_t I_b^{1-\alpha}\left[\partial_3 F\left(y(\tau), \dot{y}(\tau), {}_a^C D_\tau^\alpha[y](\tau), \tau \right) \right]$ and ${}_t I_b^{1-\alpha}\left[\partial_3 G\left(y(\tau), \dot{y}(\tau), {}_a^C D_\tau^\alpha[y](\tau), \tau \right) \right]$ are continuously differentiable on $[a, b]$.*

If $\bar{y}$ is such that

$$\partial_1 G\left(\bar{y}(t), \dot{\bar{y}}(t), {}_a^C D_t^\alpha[\bar{y}](t), t \right) - \frac{d}{dt}\left(\partial_2 G\left(\bar{y}(t), \dot{\bar{y}}(t), {}_a^C D_t^\alpha[\bar{y}](t), t \right) \right)$$

$$+ {}_t D_b^\alpha\left[\partial_3 G\left(\bar{y}(\tau), \dot{\bar{y}}(\tau), {}_a^C D_\tau^\alpha[\bar{y}](\tau), \tau \right) \right](t) \neq 0, \tag{4.28}$$

then there exists a constant λ_0 such that $\bar{y}$ satisfies

$$\partial_1 H\left(\bar{y}(t), \dot{\bar{y}}(t), {}_a^C D_t^\alpha [\bar{y}](t), t\right) - \frac{d}{dt}\left(\partial_2 H\left(\bar{y}(t), \dot{\bar{y}}(t), {}_a^C D_t^\alpha [\bar{y}](t), t\right)\right)$$

$$+ {}_t D_b^\alpha \left[\partial_3 H\left(\bar{y}(\tau), \dot{\bar{y}}(\tau), {}_a^C D_\tau^\alpha [\bar{y}](\tau), \tau\right)\right](t) = 0. \tag{4.29}$$

with $H = F - \lambda_0 G$.

Proof The result follows from Theorem 4.25.

Example 4.28 Let $\alpha \in \left(0, \frac{1}{q}\right)$ and $\xi \in \mathbb{R}$. Consider the following fractional isoperimetric problem:

$$\mathcal{I}(y) = \int_0^1 \left(\dot{y}(t) + {}_0^C D_t^\alpha [y](t)\right)^2 dt \longrightarrow \min$$

$$\mathcal{J}(y) = \int_0^1 \left(\dot{y}(t) + {}_0^C D_t^\alpha [y](t)\right) dt = \xi \tag{4.30}$$

$$y(0) = 0, \quad y(1) = \int_0^1 E_{1-\alpha}\left(-(1-\tau)^{1-\alpha}\right)\xi\, d\tau.$$

In our example (4.30), the function H of Corollary 4.27 is given by

$$H(y(t), \dot{y}(t), {}_0^C D_t^\alpha [y](t), t) = \left(\dot{y}(t) + {}_0^C D_t^\alpha [y](t)\right)^2 - \lambda_0 \left(\dot{y}(t) + {}_0^C D_t^\alpha [y](t)\right).$$

One can easily check that function

$$y(t) = \int_0^t E_{1-\alpha}\left(-(t-\tau)^{1-\alpha}\right)\xi\, d\tau \tag{4.31}$$

- is such that (4.28) holds;
- satisfies $\dot{y}(t) + {}_0^C D_t^\alpha [y](t) = \xi$ (see, e.g., p. 328, Example 5.24 (Kilbas et al. 2006)).

Moreover, (4.31) satisfies the Euler–Lagrange equation (4.29) for $\lambda_0 = 2\xi$, i.e.,

$$-\frac{d}{dt}\left(2\left(\dot{y}(t) + {}_0^C D_t^\alpha [y](t)\right) - 2\xi\right) + {}_t D_1^\alpha \left[\left(2\left(\dot{y}(\tau) + {}_0^C D_\tau^\alpha [y](\tau)\right) - 2\xi\right)\right] = 0.$$

We conclude that (4.31) is an extremal for the isoperimetric problem (4.30).

Let us consider two cases.

1. Choose $\xi = 1$. When $\alpha \to 0$ one gets from (4.30) the classical isoperimetric problem

$$\mathcal{I}(y) = \int_0^1 (\dot{y}(t) + y(t))^2 \, dt \longrightarrow \min$$

$$\mathcal{J}(y) = \int_0^1 y(t) \, dt = \frac{1}{e} \tag{4.32}$$

$$y(0) = 0 \quad y(1) = 1 - \frac{1}{e}.$$

Our extremal (4.31) is then reduced to the classical extremal $y(t) = 1 - e^{-t}$ of the isoperimetric problem (4.32).

2. Let $\alpha = \frac{1}{2}$. Then (4.30) gives the following fractional isoperimetric problem:

$$\mathcal{I}(y) = \int_0^1 \left(\dot{y}(t) + {}_0^C D_t^{\frac{1}{2}}[y](t) \right)^2 \, dt \longrightarrow \min$$

$$\mathcal{J}(y) = \int_0^1 \left(\dot{y}(t) + {}_0^C D_t^{\frac{1}{2}}[y](t) \right) \, dt = \xi \tag{4.33}$$

$$y(0) = 0 \quad y(1) = \xi \left(\text{erfc}(1)e + \frac{2}{\sqrt{\pi}} - 1 \right),$$

where erfc is the complementary error function defined by

$$\text{erfc}(z) = \frac{2}{\sqrt{\pi}} \int_z^\infty exp(-t^2) \, dt.$$

The extremal (4.31) for the particular fractional isoperimetric problem (4.33) is

$$y(t) = \xi \left(e^t \text{erfc}(\sqrt{t}) + \frac{2\sqrt{t}}{\sqrt{\pi}} - 1 \right).$$

Corollary 4.29 *Let us consider the problem of minimizing functional (4.19) subject to an isoperimetric constraint*

$$\mathcal{J}(y) = \int_a^b G \left(y(t), {}_a I_t^{1-\alpha(\cdot,\cdot)}[y](t), \dot{y}(t), {}_a^C D_t^{\alpha(\cdot,\cdot)}[y](t), t \right) \, dt = \xi \tag{4.34}$$

and the boundary conditions

$$y(a) = y_a, \quad y(b) = y_b, \tag{4.35}$$

where $\dot{y}, {}_aI_t^{1-\alpha(\cdot,\cdot)}[y], {}_a^C D_t^{\alpha(\cdot,\cdot)}[y] \in C([a,b];\mathbb{R})$ and $\alpha : \Delta \to [0, 1-\delta]$ with $\delta > \frac{1}{p}$. Moreover, we assume that

- $G \in C^1(\mathbb{R}^4 \times [a, b]; \mathbb{R})$,
- ${}_tI_b^{1-\alpha(\cdot,\cdot)}\left[\partial_2 G\left(y(\tau), {}_aI_\tau^{1-\alpha(\cdot,\cdot)}[y](\tau), \dot{y}(\tau), {}_a^C D_\tau^{\alpha(\cdot,\cdot)}[y](\tau), \tau\right)\right]$ is continuous on $[a, b]$,
- $t \mapsto \partial_3 G\left(y(t), {}_aI_t^{1-\alpha(\cdot,\cdot)}[y](t), \dot{y}(t), {}_a^C D_t^{\alpha(\cdot,\cdot)}[y](t), t\right)$

  and, ${}_tI_b^{1-\alpha(\cdot,\cdot)}\left[\partial_4 G\left(y(\tau), {}_aI_\tau^{1-\alpha(\cdot,\cdot)}[y](\tau), \dot{y}(\tau), {}_a^C D_\tau^{\alpha(\cdot,\cdot)}[y](\tau), \tau\right)\right]$ are continuously differentiable on $[a, b]$.

If $\bar{y} \in C^1([a, b]; \mathbb{R})$ is a solution to problem (4.19), (4.34) and (4.35), then there exists a real number λ_0 such that, for $H = F - \lambda_0 G$, we have

$$\partial_1 H - \frac{d}{dt}\partial_3 H + {}_tI_b^{1-\alpha(\cdot,\cdot)}[\partial_2 H] + {}_tD_b^{\alpha(\cdot,\cdot)}[\partial_4 H] = 0,$$

provided that

$$\partial_1 G - \frac{d}{dt}\partial_3 G + {}_tI_b^{1-\alpha(\cdot,\cdot)}[\partial_2 G] + {}_tD_b^{\alpha(\cdot,\cdot)}[\partial_4 G] \neq 0.$$

Here, functions $\partial_i H$ and $\partial_i G$, $i = 1, 2, 3, 4$ are evaluated at

$$\left(\bar{y}(t), {}_aI_t^{1-\alpha(\cdot,\cdot)}[\bar{y}](t), \dot{\bar{y}}, {}_a^C D_t^{\alpha(\cdot,\cdot)}[\bar{y}](t), t\right).$$

Proof The result follows from Theorem 4.25.

Theorem 4.25 can be easily extended to r subsidiary conditions of integral type. Let G_k, $k = 1, \ldots, r$, be Lagrangians of class C^1:

$$G_k : \quad \mathbb{R}^4 \times [a, b] \longrightarrow \mathbb{R}$$
$$(x_1, x_2, x_3, x_4, t) \longmapsto G_k(x_1, x_2, x_3, x_4, t).$$

and let

$$\mathcal{J}_k : A(y_a, y_b) \longrightarrow \mathbb{R} \qquad\qquad (4.36)$$
$$y \longmapsto \int_a^b G_k(y(t), K_P[y](t), \dot{y}(t), B_P[y](t), t)\, dt,$$

where $\dot{y}$ denotes the classical derivative of y, K_P is the generalized fractional integral operator with kernel belonging to $L^q(\Delta; \mathbb{R})$ and $B_P = K_P \circ \frac{d}{dt}$. Moreover, we assume that

- $K_{P*}[\partial_2 G_k(y(\tau), K_P[y](\tau), \dot{y}(\tau), B_P[y](\tau), \tau)] \in C([a, b]; \mathbb{R})$,
- $t \mapsto \partial_3 G_k(y(t), K_P[y](t), \dot{y}(t), B_P[y](t), t) \in C^1([a, b]; \mathbb{R})$,

- $K_{P^*}\left[\partial_4 G_k(y(\tau), K_P[y](\tau), \dot{y}(\tau), B_P[y](\tau), \tau)\right] \in C^1([a, b]; \mathbb{R})$.

Suppose that $\xi = (\xi_1, \ldots, \xi_r)$ and define

$$\mathcal{A}_\xi(y_a, y_b) := \{y \in \mathcal{A}(y_a, y_b) : \mathcal{J}_k[y] = \xi_k, \quad k = 1, \ldots, r\}.$$

Next theorem gives necessary optimality condition for a minimizer of functional $\mathcal{I}$ subject to r isoperimetric constraints.

Theorem 4.30 *Let $\bar{y}$ be a minimizer of $\mathcal{I}$ in the class $\mathcal{A}_\xi(y_a, y_b)$. If one can find functions $\eta_1, \ldots, \eta_r \in \mathcal{A}(0, 0)$ such that the matrix $A = (a_{kl})$, with*

$$a_{kl} := \int_a^b \left(\partial_1 G_k(\star_{\bar{y}})(t) + K_{P^*}\left[\partial_2 G_k(\star_{\bar{y}})(\tau)\right](t)\right) \cdot \eta_l(\tau)$$

$$+ \left(\partial_3 G_k(\star_{\bar{y}})(t) + K_{P^*}\left[\partial_4 G_k(\star_{\bar{y}})(\tau)\right](t)\right) \cdot \dot{\eta}_l(t) \, dt,$$

has rank equal to r, then there exist $\lambda_1, \ldots, \lambda_r \in \mathbb{R}$ such that, for $H = F - \sum_{k=1}^{r} \lambda_k G_k$, minimizer $\bar{y}$ satisfies

$$\frac{d}{dt}\left[\partial_3 H(\star_{\bar{y}})(t)\right] + A_{P^*}\left[\partial_4 H(\star_{\bar{y}})(\tau)\right](t)$$

$$= \partial_1 H(\star_{\bar{y}})(t) + K_{P^*}\left[\partial_2 H(\star_{\bar{y}})(\tau)\right](t), \quad t \in (a, b), \qquad (4.37)$$

where $(\star_{\bar{y}})(t) = (\bar{y}(t), K_P[\bar{y}](t), \dot{\bar{y}}(t), B_P[\bar{y}](t), t)$.

Proof Let us define

$$\phi : [-\varepsilon_0, \varepsilon_0] \times [-\varepsilon_1, \varepsilon_1] \times \cdots \times [-\varepsilon_r, \varepsilon_r] \longrightarrow \mathbb{R}$$
$$(h_0, h_1, \ldots, h_r) \longmapsto \mathcal{I}(\bar{y} + h_0\eta_0 + h_1\eta_1 + \cdots + h_r\eta_r)$$

and

$$\psi_k : [-\varepsilon_0, \varepsilon_0] \times [-\varepsilon_1, \varepsilon_1] \times \cdots \times [-\varepsilon_r, \varepsilon_r] \to \mathbb{R}$$

$$(h_0, h_1, \ldots, h_r) \mapsto \mathcal{J}_k(\bar{y} + h_0\eta_0 + h_1\eta_1 + \cdots + h_r\eta_r) - \xi_k.$$

Observe that ϕ, ψ_k are functions of class $C^1\left([-\varepsilon_0.\varepsilon_0] \times \cdots \times [-\varepsilon_r, \varepsilon_r]; \mathbb{R}\right)$, $A = \left(\frac{\partial \psi_k}{\partial h_l}\Big|_0\right)$ and that $\psi_k(0, 0, \ldots, 0) = 0$. Moreover, because $\bar{y}$ is a minimizer of functional $\mathcal{I}$, we have

$$\phi(0, \ldots, 0) \leq \phi(h_0, h_1, \ldots, h_r).$$

From the classical multiplier theorem, there are $\lambda_1, \ldots, \lambda_r \in \mathbb{R}$ such that

$$\nabla \phi_l(0, \ldots, 0) + \sum_{k=1}^{r} \lambda_k \nabla \psi_k(0, \ldots, 0) = 0. \qquad (4.38)$$

From (4.38), we can compute $\lambda_1, \ldots, \lambda_r$, turning out to be independent of the choice of $\eta_0 \in \mathcal{A}(0, 0)$. Finally, by the fundamental lemma of the calculus of variations, we arrive to (4.37).

4.5 Noether's Theorem

Emmy Noether's classical work (Noether 1918) states that a conservation law in variational mechanics follows whenever the Lagrangian function is invariant under a one-parameter continuous group of transformations, that transform dependent and/or independent variables. This result not only unifies conservation laws but also suggests a way to discover new ones. In this section we consider variational problems that depend on generalized fractional integrals and derivatives. Following the methods used in Cresson (2007), Frederico and Torres (2008, 2010), Jost and Li-Jost (1998), we apply Euler–Lagrange equations to formulate a generalized fractional version of Noether's theorem without transformation of time. We start by introducing the notions of generalized fractional extremal and one-parameter family of infinitesimal transformations.

Definition 4.31 A function $y \in C^1([a, b]; \mathbb{R})$ that is a solution to Eq. (4.11) with $K_P[y], B_P[y] \in C([a, b]; \mathbb{R})$, is said to be a generalized fractional extremal.

We consider a one-parameter family of transformations of the form $\hat{y} = \phi(\theta, t, y)$, where ϕ is a map of class C^2:

$$\phi : [-\varepsilon, \varepsilon] \times [a, b] \times \mathbb{R} \longrightarrow \mathbb{R}$$
$$(\theta, t, y) \longmapsto \phi(\theta, t, y),$$

such that $\phi(0, t, y) = y$. Note that, using Taylor's expansion of $\phi(\theta, t, y)$ in a neighborhood of 0, one has

$$\hat{y} = \phi(0, t, y) + \theta \frac{\partial}{\partial \theta} \phi(0, t, y) + o(\theta),$$

provided that $\theta \in [-\varepsilon, \varepsilon]$. Moreover, having in mind that $\phi(0, t, y) = y$ and denoting $\xi(t, y) = \frac{\partial}{\partial \theta} \phi(0, t, y)$, one has

$$\hat{y} = y + \theta \xi(t, y) + o(\theta), \qquad (4.39)$$

so that, for $\theta \in [-\varepsilon, \varepsilon]$, the linear approximation to the transformation is

$$\hat{y} \approx y + \theta \xi(t, y).$$

Let $y : [a, b] \rightarrow \mathbb{R}$ be given by $y = y(t)$. Then, for sufficiently small θ, the transformation (4.39) carries the curve $y = y(t)$ into a family of curves $\hat{y} = \hat{y}(t) = \phi(\theta, t, y(t))$. Now, we introduce the notion of invariance.

Definition 4.32 We say that the Lagrangian F is invariant under the one-parameter family of infinitesimal transformations (4.39), where ξ is such that $t \mapsto \xi(t, y(t)) \in C^1([a, b]; \mathbb{R})$ with $K_P[\tau \mapsto \xi(\tau, y(\tau))]$, $B_P[\tau \mapsto \xi(\tau, y(\tau))] \in C([a, b]; \mathbb{R})$ if

$$F\left(y(t), K_P[y](t), \dot{y}(t), B_P[y](t), t\right) = F\left(\hat{y}(t), K_P[\hat{y}](t), \dot{\hat{y}}(t), B_P[\hat{y}](t), t\right), \tag{4.40}$$

for all $\theta \in [-\varepsilon, \varepsilon]$, and all $y \in C^1([a, b]; \mathbb{R})$ with $K_P[y], B_P[y] \in C([a, b]; \mathbb{R})$.

In order to state Noether's theorem in a compact form, we introduce the following two bilinear operators:

$$\mathbf{D}[f, g] := f \cdot A_{P^*}[g] + g \cdot B_P[f], \tag{4.41}$$

$$\mathbf{I}[f, g] := -f \cdot K_{P^*}[g] + g \cdot K_P[f]. \tag{4.42}$$

Theorem 4.33 (Generalized Fractional Noether's Theorem) *Let F be invariant under the one-parameter family of infinitesimal transformations (4.39). Then, for every generalized fractional extremal, the following equality holds:*

$$\frac{d}{dt}\left(\xi(t, y(t)) \cdot \partial_3 F(\star_y)(t)\right) + \mathbf{D}\left[\xi(t, y(t)), \partial_4 F(\star_y)(t)\right]$$

$$+ \mathbf{I}\left[\xi(t, y(t)), \partial_2 F(\star_y)(t)\right] = 0, \quad t \in (a, b), \tag{4.43}$$

where $(\star_y)(t) = (y(t), K_P[y](t), \dot{y}(t), B_P[y](t), t)$.

Proof By Eq. (4.40) one has

$$\frac{d}{d\theta}\left[F\left(\hat{y}(t), K_P[\hat{y}](t), \dot{\hat{y}}(t), B_P[\hat{y}](t), t\right)\right]\Bigg|_{\theta=0} = 0.$$

The usual chain rule implies

$$\partial_1 F(\star_{\hat{y}})(t) \cdot \frac{d}{d\theta}\hat{y}(t) + \partial_2 F(\star_{\hat{y}})(t) \cdot \frac{d}{d\theta}K_P[\hat{y}](t)$$

$$+ \partial_3 F(\star_{\hat{y}})(t) \cdot \frac{d}{d\theta}\dot{\hat{y}}(t) + \partial_4 F(\star_{\hat{y}})(t) \cdot \frac{d}{d\theta}B_P[\hat{y}](t)\Bigg|_{\theta=0} = 0.$$

By linearity of K_P, B_P differentiating with respect to θ, and putting $\theta = 0$, we have

$$\partial_1 F(\star_y)(t) \cdot \xi(t, y(t)) + \partial_2 F(\star_y)(t) \cdot K_P[\tau \mapsto \xi(\tau, y(\tau))](t)$$

$$+ \partial_3 F(\star_y)(t) \cdot \frac{d}{dt}\xi(t, y(t)) + \partial_4 F(\star_y)(t) \cdot B_P[\tau \mapsto \xi(\tau, y(\tau))](t) = 0.$$

Now, using the generalized Euler–Lagrange equation (4.11) and formulas (4.41) and (4.42), one arrives to (4.43).

Example 4.34 Let $P = \langle a, t, b, \lambda, \mu \rangle$ and $y \in C^1([a, b]; \mathbb{R})$ with $B_P[y] \in C([a, b]; \mathbb{R})$. Consider Lagrangian $F(B_P[y](t), t)$ and transformations

$$\hat{y}(t) = y(t) + \varepsilon c + o(\varepsilon), \qquad (4.44)$$

where c is a constant. Then, we have

$$F(B_P[y](t), t) = F(B_P[\hat{y}](t), t).$$

Therefore, F is invariant under (4.44) and the generalized fractional Noether's theorem indicates that

$$A_{P*}[\partial_1 F(B_P[y](\tau), \tau)](t) = 0, \quad t \in (a, b), \qquad (4.45)$$

along any generalized fractional extremal y. Notice that Eq. (4.45) can be written in the form

$$\frac{d}{dt}(K_{P*}[\partial_1 F(B_P[y](\tau), \tau)](t)) = 0, \qquad (4.46)$$

that is, quantity $K_{P*}[\partial_1 F(B_P[y](\tau), \tau)]$ is conserved along all generalized fractional extremals and this quantity, following the classical approach, can be called a generalized fractional constant of motion.

Similarly to previous sections, one can obtain from Theorem 4.33 results regarding to constant and variable order fractional integrals and derivatives.

Corollary 4.35 *If for any $y \in C^1([a, b]; \mathbb{R})$ the following equality is satisfied*

$$F\left(y(t), \dot{y}(t), \lambda \, {}_a^C D_t^\alpha[y](t) + \mu \, {}_t^C D_b^\alpha[y](t), t\right)$$

$$= F\left(\hat{y}(t), \dot{\hat{y}}(t), \lambda \, {}_a^C D_t^\alpha[\hat{y}](t) + \mu \, {}_t^C D_b^\alpha[\hat{y}](t), t\right),$$

where $\hat{y}$ is the family (4.39), then we have

$$\frac{d}{dt}\left(\xi(t,y(t))\cdot\partial_2 F\left(y(t),\dot{y}(t),\lambda\,{}_a^C D_t^\alpha[y](t)+\mu\,{}_t^C D_b^\alpha[y](t),t\right)\right)$$

$$-\xi(t,y(t))\cdot\left(\lambda_t D_b^\alpha\left[\partial_3 F\left(y(\tau),\dot{y}(\tau),\lambda\,{}_a^C D_\tau^\alpha[y](\tau)+\mu\,{}_\tau^C D_b^\alpha[y](\tau),\tau\right)\right](t)\right.$$

$$+\mu\,{}_a D_t^\alpha\left[\partial_3 F\left(y(\tau),\dot{y}(\tau),\lambda\,{}_a^C D_\tau^\alpha[y](\tau)+\mu\,{}_\tau^C D_b^\alpha[y](\tau),\tau\right)\right](t)\right)$$

$$+\partial_3 F\left(y(t),\dot{y}(t),\lambda\,{}_a^C D_t^\alpha[y](t)+\mu\,{}_t^C D_b^\alpha[y](t),t\right)$$

$$\cdot\left(\lambda\,{}_a^C D_t^\alpha[\xi(\tau,y(\tau))](t)+\mu\,{}_t^C D_b^\alpha[\xi(\tau,y(\tau))](t)\right)=0$$

along all solutions of the Euler–Lagrange equation (4.18).

Proof Corollary 4.35 is a simple consequence of Theorem 4.33.

Corollary 4.36 *Let* $y\in C^1([a,b];\mathbb{R})$ *with* ${}_a I_t^{1-\alpha(\cdot,\cdot)}[y]$, ${}_a^C D_t^{\alpha(\cdot,\cdot)}[y]\in C([a,b];\mathbb{R})$ *and suppose that*

$$F\left(y(t),{}_a I_t^{1-\alpha(\cdot,\cdot)}[y](t),\dot{y}(t),{}_a^C D_t^{\alpha(\cdot,\cdot)}[y](t),t\right)$$

$$=F\left(\hat{y}(t),{}_a I_t^{1-\alpha(\cdot,\cdot)}[\hat{y}](t),\dot{\hat{y}}(t),{}_a^C D_t^{\alpha(\cdot,\cdot)}[\hat{y}](t),t\right)$$

where $\hat{y}$ *is the family* (4.39) *such that* $t\mapsto\xi(t,y(t))\in C^1([a,b];\mathbb{R})$ *and* ${}_a I_t^{1-\alpha(\cdot,\cdot)}[\tau\mapsto\xi(\tau,y(\tau))]$, ${}_a^C D_t^{\alpha(\cdot,\cdot)}[\tau\mapsto\xi(\tau,y(\tau))]\in C([a,b];\mathbb{R})$. *Then all solutions of the Euler–Lagrange equation* (4.21) *must satisfy*

$$\frac{d}{dt}\left(\xi(t,y(t))\cdot\partial_3 F\left(y(t),{}_a I_t^{1-\alpha(\cdot,\cdot)}[y](t),\dot{y}(t),{}_a^C D_t^{\alpha(\cdot,\cdot)}[y](t),t\right)\right)$$

$$-\xi(t,y(t))\cdot{}_t D_b^{\alpha(\cdot,\cdot)}\left[\partial_4 F\left(y(\tau),{}_a I_\tau^{1-\alpha(\cdot,\cdot)}[y](\tau),\dot{y}(\tau),{}_a^C D_\tau^{\alpha(\cdot,\cdot)}[y](\tau),\tau\right)\right](t)$$

$$+\partial_4 F\left(y(t),{}_a I_t^{1-\alpha(\cdot,\cdot)}[y](t),\dot{y}(t),{}_a^C D_t^{\alpha(\cdot,\cdot)}[y](t),t\right)\cdot{}_a^C D_t^{\alpha(\cdot,\cdot)}[\xi(\tau,y(\tau))](t)$$

$$-\xi(t,y(t))\cdot{}_t I_b^{1-\alpha(\cdot,\cdot)}\left[\partial_2 F\left(y(\tau),{}_a I_\tau^{1-\alpha(\cdot,\cdot)}[y](\tau),\dot{y}(\tau),{}_a^C D_\tau^{\alpha(\cdot,\cdot)}[y](\tau),\tau\right)\right](t)$$

$$+\partial_2 F\left(y(t),{}_a I_t^{1-\alpha(\cdot,\cdot)}[y](t),\dot{y}(t),{}_a^C D_t^{\alpha(\cdot,\cdot)}[y](t),t\right)$$

$$\cdot{}_a I_t^{1-\alpha(\cdot,\cdot)}[\xi(\tau,y(\tau))](t)=0,$$

$t\in(a,b)$.

Proof Corollary 4.36 can be easily obtained from Theorem 4.33.

Corollary 4.37 *Suppose that* $y \in C^1([a, b]; \mathbb{R})$ *and for family (4.39) one has*

$$F(y(t), {}_aI_t^{1-\alpha}[y](t), \dot{y}(t), {}_a^C D_t^\alpha[y](t), t)\,dt$$
$$= F(\hat{y}(t), {}_aI_t^{1-\alpha}[\hat{y}](t), \dot{\hat{y}}(t), {}_a^C D_t^\alpha[\hat{y}](t), t).$$

Then

$$\frac{d}{dt}\left(\xi(t, y(t)) \cdot \partial_2 F(y(t), {}_aI_t^{1-\alpha}[y](t), \dot{y}(t), {}_a^C D_t^\alpha[y](t), t)\right)$$

$$- \xi(t, y(t)) \cdot {}_tD_b^\alpha[\partial_4 F(y(\tau), {}_aI_\tau^{1-\alpha}[y](\tau), \dot{y}(\tau), {}_a^C D_\tau^\alpha[y](\tau), \tau)]$$

$$+ \partial_4 F(y(t), {}_aI_t^{1-\alpha}[y](t), \dot{y}(t), {}_a^C D_t^\alpha[y](t), t) \cdot {}_a^C D_t^\alpha[\xi(\tau, y(\tau))]$$

$$- \xi(t, y(t)) \cdot {}_tI_b^{1-\alpha}[\partial_2 F(y(\tau), {}_aI_\tau^{1-\alpha}[y](\tau), \dot{y}(\tau), {}_a^C D_\tau^\alpha[y](\tau), \tau)]$$

$$+ \partial_2 F(y(t), {}_aI_t^{1-\alpha}[y](t), \dot{y}(t), {}_a^C D_t^\alpha[y](t), t) \cdot {}_aI_t^{1-\alpha}[\xi(\tau, y(\tau))] = 0$$

along any solution of the Euler–Lagrange equation (4.13).

Proof Corollary 4.37 can be easily obtained from Theorem 4.33.

4.6 Variational Calculus in Terms of a Generalized Integral

In this section, we develop a generalized fractional calculus of variations, by considering very general problems, where the classical integrals are substituted by generalized fractional integrals, and the Lagrangians depend not only on classical derivatives but also on generalized fractional operators. By choosing particular operators and kernels, one obtains some results available in the literature of mathematical physics (Herrera et al. 1986).

Let $R = \langle a, b, b, 1, 0\rangle$, $P = \langle a, t, b, \lambda, \mu\rangle$ and consider the problem of finding a function $\bar{y}$ that gives minimum value to the functional

$$\mathcal{I} : \mathcal{A}(y_a, y_b) \longrightarrow \mathbb{R} \tag{4.47}$$
$$y \longmapsto K_R\left[F(y(t), K_P[y](t), \dot{y}(t), B_P[y](t), t)\right](b),$$

where K_R and K_P are generalized fractional integrals with kernels $k(x, t)$ and $h(t, \tau)$, respectively, being elements of $L^q(\Delta; \mathbb{R})$, $B_P = K_P \circ \frac{d}{dt}$ and F is a Lagrangian of class C^1:

$$F : \quad \mathbb{R}^4 \times [a, b] \longrightarrow \mathbb{R}$$
$$(x_1, x_2, x_3, x_4, t) \longmapsto F(x_1, x_2, x_3, x_4, t).$$

Moreover, we assume that

- $t \mapsto k(b, t) \cdot \partial_1 F(y(t), K_P[y](t), \dot{y}(t), B_P[y](t), t) \in C([a, b]; \mathbb{R})$,
- $K_{P^*}[k(b, \tau)\partial_2 F(y(\tau), K_P[y](\tau), \dot{y}(\tau), B_P[y](\tau), \tau)] \in C([a, b]; \mathbb{R})$,
- $t \mapsto k(b, t) \cdot \partial_3 F(y(t), K_P[y](t), \dot{y}(t), B_P[y](t), t) \in C^1([a, b]; \mathbb{R})$,
- $K_{P^*}[k(b, \tau)\partial_4 F(y(\tau), K_P[y](\tau), \dot{y}(\tau), B_P[y](\tau), \tau)] \in C^1([a, b]; \mathbb{R})$.

Theorem 4.38 *If $\bar{y} \in \mathcal{A}(y_a, y_b)$ is a minimizer of functional (4.47), then $\bar{y}$ satisfies the generalized Euler–Lagrange equation*

$$k(b, t) \cdot \partial_1 F(\star_y)(t) - \frac{d}{dt}\left(\partial_3 F(\star_y)(t) \cdot k(b, t)\right)$$
$$- A_{P^*}\left[k(b, \tau) \cdot \partial_4 F(\star_y)(\tau)\right](t)$$
$$+ K_{P^*}\left[k(b, \tau) \cdot \partial_2 F(\star_y)(\tau)\right](t) = 0, \quad t \in (a, b), \qquad (4.48)$$

where $(\star_y) = (y(t), K_P[y](t), \dot{y}(t), B_P[y](t), t)$.

Proof One can prove Theorem 4.38 in a similar way as Theorem 4.10. $\square$

Example 4.39 Let $R = \langle 0, 1, 1, 1, 0 \rangle$, and $P = \langle 0, t, 1, 1, 0 \rangle$. Consider the following problem:

$$\mathcal{J}(y) = K_R\left[t K_P[y](t) + \sqrt{1 - (K_P[y](t))^2}\right](1) \longrightarrow \min,$$

$$y(0) = 1, \quad y(1) = \frac{\sqrt{2}}{4} + \int_0^1 u(1 - \tau)\frac{1}{(1 + \tau^2)^{\frac{3}{2}}}\, d\tau,$$

with kernel h such that $h(t, \tau) = l(t - \tau)$, $l \in C^1([0, 1]; \mathbb{R})$, $l(t) \equiv 0$ for $t < 0$, $l(0) = 1$, and we can find $M > 0$ and $\rho > 0$ such that $|l(t)| \leq Me^{\rho t}$, $t \geq 0$. Here, the resolvent $u(t)$ is related to the kernel $l(t)$ by $u(t) = \mathcal{L}^{-1}\left[\frac{1}{s\tilde{l}(s)} - 1\right](t)$, $\tilde{l}(s) = \mathcal{L}[l(t)](s)$, where $\mathcal{L}$ and $\mathcal{L}^{-1}$ are the direct and the inverse Laplace operators, respectively. We apply Theorem 4.38 with Lagrangian F given by $F(x_1, x_2, x_3, x_4, t) = tx_2 + \sqrt{1 - x_2^2}$. Because

$$y(t) = \frac{1}{(1 + t^2)^{\frac{3}{2}}} + \int_0^t u(t - \tau)\frac{1}{(1 + \tau^2)^{\frac{3}{2}}}\, d\tau$$

is the solution to the Volterra integral equation of first kind (see, e.g., Eq. 16, p. 114 of Polyanin and Manzhirov (1998))

$$K_P[y](t) = \frac{t\sqrt{1 + t^2}}{1 + t^2},$$

it satisfies our generalized Euler–Lagrange equation (4.48), that is,

$$K_{P*}\left[k(1,\tau)\left(\frac{-K_P[y](\tau)}{\sqrt{1-(K_P[y](\tau))^2}}+\tau\right)\right](t)=0.$$

In particular, for the kernel $l^\beta(t-\tau)=\cosh(\beta(t-\tau))$, the boundary conditions are $y(0)=1$ and $y(1)=1+\beta^2(1-\sqrt{2})$, and the solution is $y(t)=\dfrac{1}{(1+t^2)^{\frac{3}{2}}}+\beta^2\left(1-\sqrt{1+t^2}\right)$ (cf. p. 22 in Polyanin and Manzhirov (1998)).

Follows a direct corollary of Theorem 4.38.

Corollary 4.40 *If $\bar{y}\in C^1\left([a,b];\mathbb{R}\right)$ is a solution to the problem of minimizing*

$$\mathcal{J}(y)={_a}I_b^\alpha\left[F\left(y(t),\dot{y}(t),{_a^C}D_t^\beta\,[y]\,(t),t\right)\right](b),$$

subject to the boundary conditions

$$y(a)=y_a,\quad y(b)=y_b,$$

where $\alpha,\beta\in(0,\frac{1}{q})$, $F\in C^2\left(\mathbb{R}^3\times[a,b];\mathbb{R}\right)$, then

$$\partial_1 F\left(\bar{y}(t),\dot{\bar{y}}(t),{_a^C}D_t^\beta\,[\bar{y}]\,(t),t\right)\cdot(b-t)^{\alpha-1}$$

$$-\frac{d}{dt}\left(\partial_2 F\left(\bar{y}(t),\dot{\bar{y}}(t),{_a^C}D_t^\beta\,[\bar{y}]\,(t),t\right)\cdot(b-t)^{\alpha-1}\right)$$

$$+{_t}D_b^\beta\left[(b-\tau)^{\alpha-1}\cdot\partial_3 F\left(\bar{y}(\tau),\dot{\bar{y}}(\tau),{_a^C}D_\tau^\beta\,[\bar{y}]\,(\tau),\tau\right)\right]=0,\ t\in(a,b).$$

If the Lagrangian of functional (4.47) does not depend on generalized fractional operators B and K, then Theorem 4.38 gives the following result: if $y\in C^1([a,b];\mathbb{R})$ is a solution to the problem of extremizing

$$\mathcal{I}(y)=\int_a^b F\left(y(t),\dot{y}(t),t\right)k(b,t)\,dt \tag{4.49}$$

subject to $y(a)=y_a$ and $y(b)=y_b$, then

$$\partial_1 F\left(y(t),\dot{y}(t),t\right)-\frac{d}{dt}\partial_2 F\left(y(t),\dot{y}(t),t\right)$$

$$=\frac{1}{k(b,t)}\cdot\left(\frac{d}{dt}k(b,t)\right)\partial_2 F\left(y(t),\dot{y}(t),t\right). \tag{4.50}$$

We recognize on the right-hand side of (4.50) the generalized weak dissipative parameter

$$\delta(t) = \frac{1}{k(b, t)} \cdot \frac{d}{dt} k(b, t).$$

Now, let us consider an example of Lagrangian associated to functional (4.49).

Example 4.41 Let us consider kernel $k^\alpha(b, t) = e^{\alpha(b-t)}$ and the Lagrangian

$$L(y(t), \dot{y}(t), t) = \frac{1}{2} m \dot{y}^2(t) - V(y(t)),$$

where $V(y)$ is the potential energy and m stands for mass. The Euler–Lagrange equation (4.50) gives the following second-order ordinary differential equation:

$$\ddot{y}(t) - \alpha \dot{y}(t) = -\frac{1}{m} V'(y(t)). \tag{4.51}$$

Equation (4.51) coincides with (14) of Herrera et al. (1986), obtained by modification of Hamilton's principle.

Example 4.42 Let us consider the Caldirola–Kanai Lagrangian

$$L(y(t), \dot{y}(t), t) = m(t) \left(\frac{\dot{y}^2(t)}{2} - \omega^2 \frac{y^2(t)}{2} \right), \tag{4.52}$$

which describes a dynamical oscillatory system with exponentially increasing time-dependent mass, where ω is the frequency and $m(t) = m_0 e^{-\gamma b} e^{\gamma t} = \bar{m}_0 e^{\gamma t}$, $\bar{m}_0 = m_0 e^{-\gamma b}$. Using our generalized Euler–Lagrange equation (4.50) with kernel $k(b, t)$ to Lagrangian (4.52), we obtain

$$\ddot{y}(t) + (\delta(t) + \gamma) \dot{y}(t) + \omega^2 y(t) = 0.$$

4.7 Generalized Variational Calculus of Several Variables

Variational problems with functionals depending on several variables arise, for example, in mechanics problems, which involve systems with infinite number of degrees of freedom, like a vibrating elastic solid. Fractional variational problems involving multiple integrals have been already studied in different contexts. We can mention here Almeida et al. (2010), Baleanu and Muslih (2005), Cresson (2007), Odzijewicz and Torres (2011), where the multidimensional fractional Euler–Lagrange equations for a field were obtained, or Malinowska (2012, 2013) where first and second fractional Noether-type theorems are proved. In this section, we present a more general approach to the subject by considering functionals depending on generalized fractional operators.

4.7.1 Multidimensional Generalized Fractional Integration by Parts

In this section, it is of our interest to obtain integration by parts formula for generalized fractional operators. We shall denote a point in Ω_n by $t = (t_1, \ldots, t_n)$, where $\Omega_n = (a_1, b_1) \times \cdots \times (a_n, b_n)$, and by $dt = dt_1, \ldots, dt_n$. Throughout this subsection $i \in \{1, \ldots, n\}$ is arbitrary but fixed.

Theorem 4.43 *Let $P_i = \langle a_i, t_i, b_i, \lambda_i, \mu_i \rangle$ be the parameter set and let K_{P_i} be the generalized partial fractional integral with k_i being a difference kernel such that $k_i \in L^1(0, b_i - a_i; \mathbb{R})$. If $f : \mathbb{R}^n \to \mathbb{R}$ and $\eta : \mathbb{R}^n \to \mathbb{R}$, $f, \eta \in C\left(\bar{\Omega}_n; \mathbb{R}\right)$, then the generalized partial fractional integrals satisfy the following identity:*

$$
\int_{\Omega_n} f(t) \cdot K_{P_i}[\eta](t) \, dt = \int_{\Omega_n} \eta(t) \cdot K_{P_i^*}[f](t) \, dt,
$$

where P_i^* is the dual of P_i.

Proof Let $P_i = \langle a_i, t_i, b_i, \lambda_i, \mu_i \rangle$ and $f, \eta, \in C\left(\bar{\Omega}_n; \mathbb{R}\right)$. Since f and η, are continuous functions on $\bar{\Omega}_n$, they are bounded on $\bar{\Omega}_n$, i.e., there exist real numbers $C, D > 0$ such that $|f(t)| \leq C$, $|\eta(t)| \leq D$, for all $t \in \bar{\Omega}_n$. Therefore,

$$
\int_{\Omega_n} \left(\int_{a_i}^{t_i} |\lambda_i k_i(t_i - \tau)| \cdot |f(t)| \cdot |\eta(t_1, \ldots, t_{i-1}, \tau, t_{i+1}, \ldots, t_n)| \, d\tau \right.
$$

$$
\left. + \int_{t_i}^{b_i} |\mu_i k_i(\tau - t_i)| \cdot |f(t)| \cdot |\eta(t_1, \ldots, t_{i-1}, \tau, t_{i+1}, \ldots, t_n)| \, d\tau \right) dt
$$

$$
\leq C \cdot D \int_{\Omega_n} \left(\int_{a_i}^{t_i} |\lambda_i k_i(t_i - \tau)| \, d\tau + \int_{t_i}^{b_i} |\mu_i k_i(\tau - t_i)| \, d\tau \right) dt
$$

$$
\leq C \cdot D \left(|\mu_i| + |\lambda_i| \right) \|k_i\|_{L^1(0, b_i - a_i; \mathbb{R})} \int_{\Omega_n} dt
$$

$$
= C \cdot D \left(|\mu_i| + |\lambda_i| \right) \|k_i\|_{L^1(0, b_i - a_i; \mathbb{R})} \cdot \prod_{i=1}^{n} (b_i - a_i) < \infty.
$$

Hence, we can use Fubini's theorem to change the order of integration in the iterated integrals:

$$\int\limits_{\Omega_n} f(t) \cdot K_{P_i}[\eta](t)\, dt_n, \ldots, dt_1$$

$$= \int\limits_{\Omega_n} \left(\lambda_i \int\limits_{a_i}^{t_i} f(t) k_i (t_i - \tau) \eta(t_1, \ldots, t_{i-1}, \tau, t_{i+1}, \ldots, t_n)\, d\tau \right.$$

$$\left. + \mu_i \int\limits_{t_i}^{b_i} f(t) k_i (\tau - t_i) \eta(t_1, \ldots, t_{i-1}, \tau, t_{i+1}, \ldots, t_n)\, d\tau \right) dt_n, \ldots, dt_1$$

$$= \int\limits_{\Omega_n} \left(\lambda_i \int\limits_{\tau}^{b_i} f(t) k_i (t_i - \tau) \eta(t_1, \ldots, t_{i-1}, \tau, t_{i+1}, \ldots, t_n)\, dt_i \right.$$

$$\left. + \mu_i \int\limits_{a_i}^{\tau} f(t) k_i (\tau - t_i) \eta(t_1, \ldots, t_{i-1}, \tau, t_{i+1}, \ldots, t_n)\, dt_i \right)$$

$$\times dt_n, \ldots, dt_{i-1}\, d\tau\, dt_{i+1}, \ldots, dt_1$$

$$= \int\limits_{\Omega_n} \eta(t_1, \ldots, t_{i-1}, \tau, t_{i+1}, \ldots, t_n) \left(\lambda_i \int\limits_{\tau}^{b_i} f(t) k_i (t_i - \tau)\, dt_i \right.$$

$$\left. + \mu_i \int\limits_{a_i}^{\tau} f(t) k_i (\tau - t_i)\, dt_i \right) dt_n, \ldots, dt_{i-1}\, d\tau\, dt_{i+1}, \ldots, dt_1$$

$$= \int\limits_{\Omega_n} \eta(t) \cdot K_{P_i^*}[f](t)\, dt_n, \ldots, dt_1.$$

As corollaries, we obtain the following integration by parts formulas for constant and variable order fractional integrals.

Corollary 4.44 *Let* $0 < \alpha_i < 1$, *and let* $f : \mathbb{R}^n \to \mathbb{R}$, $\eta : \mathbb{R}^n \to \mathbb{R}$ *be such that* $f, \eta \in C(\bar{\Omega}_n; \mathbb{R})$. *Then the following formula holds:*

$$\int\limits_{\Omega_n} f(t) \cdot {}_{a_i} I_{t_i}^{\alpha_i}[\eta](t)\, dt = \int\limits_{\Omega_n} \eta(t) \cdot {}_{t_i} I_{b_i}^{\alpha_i}[f](t)\, dt. \tag{4.53}$$

Corollary 4.45 *Suppose that* $\alpha : [0, b_i - a_i] \to [0, 1]$, *and that* $f : \mathbb{R}^n \to \mathbb{R}$, $\eta : \mathbb{R}^n \to \mathbb{R}$ *are such that* $f, \eta \in C(\bar{\Omega}_n; \mathbb{R})$. *Then,*

$$\int\limits_{\Omega_n} f(t) \cdot {}_{a_i} I_{t_i}^{\alpha_i(\cdot)}[\eta](t)\, dt = \int\limits_{\Omega_n} \eta(t) \cdot {}_{t_i} I_{b_i}^{\alpha_i(\cdot)}[f](t)\, dt. \tag{4.54}$$

Theorem 4.46 (Generalized Fractional Integration by Parts for Several Variables)
*Let $P_i = \langle a_i, t_i, b_i, \lambda_i, \mu_i \rangle$ be the parameter set and $f, \eta \in C^1\left(\bar{\Omega}_n; \mathbb{R}\right)$. Moreover,
let $B_{P_i} = \frac{d}{dt} \circ K_{P_i}$, where K_{P_i} is the generalized partial fractional integral with
difference kernel, i.e., $k_i = k_i(t_i - \tau)$ such that $k_i \in L_1(0, b_i - a_i; \mathbb{R})$, and $K_{P_i^*}[f] \in
C^1\left(\bar{\Omega}_n; \mathbb{R}\right)$. Then*

$$\int_{\Omega_n} f(t) \cdot B_{P_i}[\eta](t)\, dt = \int_{\partial\Omega_n} \eta(t) \cdot K_{P_i^*}[f](t) \cdot \nu^i\, d(\partial\Omega_n) - \int_{\Omega_n} \eta(t) \cdot A_{P_i^*}[f](t)\, dt,$$

where ν^i is the outward pointing unit normal to $\partial\Omega_n$.

Proof By the definition of generalized partial Caputo fractional derivative,
Theorem 4.43, and the standard integration by parts formula (see, e.g., (Evans 2010))
one has

$$\int_{\Omega_n} f(t) \cdot B_{P_i}[\eta](t)\, dt = \int_{\Omega_n} f(t) \cdot K_{P_i}\left[\frac{\partial\eta}{\partial t_i}\right](t)\, dt = \int_{\Omega_n} \frac{\partial\eta(t)}{\partial t_i} \cdot K_{P_i^*}[f](t)\, dt$$

$$= \int_{\partial\Omega_n} \eta(t) \cdot K_{P_i^*}[f](t) \cdot \nu^i\, d(\partial\Omega_n) - \int_{\Omega_n} \eta(t) \cdot \frac{\partial}{\partial t_i} K_{P_i^*}[f](t)\, dt$$

$$= \int_{\partial\Omega_n} \eta(t) \cdot K_{P_i^*}[f](t) \cdot \nu^i\, d(\partial\Omega_n) - \int_{\Omega_n} \eta(t) \cdot A_{P_i^*}[f](t)\, dt.$$

Next corollaries present multidimensional integration by parts formulas for constant and variable order fractional derivatives.

Corollary 4.47 *If $0 < \alpha_i < 1$, functions $f : \mathbb{R}^n \to \mathbb{R}$ and $\eta : \mathbb{R}^n \to \mathbb{R}$ are such
that $f, \eta \in C^1(\bar{\Omega}_n; \mathbb{R})$ and $_t I_{b_i}^{1-\alpha_i}[f] \in C^1(\bar{\Omega}_n; \mathbb{R})$, then*

$$\int_{\Omega_n} f(t) \cdot {}_{a_i}^C D_{t_i}^{\alpha_i}[\eta](t)\, dt = \int_{\partial\Omega_n} \eta(t) \cdot {}_t I_{b_i}^{1-\alpha_i}[f](t) \cdot \nu^i\, d(\partial\Omega_n)$$

$$+ \int_{\Omega_n} \eta(t) \cdot {}_{t_i} D_{b_i}^{\alpha_i}[f](t)\, dt.$$

Corollary 4.48 *If $\alpha_i : [0, b_i - a_i] \to [0, 1]$, functions $f : \mathbb{R}^n \to \mathbb{R}$ and $\eta : \mathbb{R}^n \to \mathbb{R}$
are such that $f, \eta \in C^1(\bar{\Omega}_n; \mathbb{R})$ and $_{t_i} I_{b_i}^{\alpha_i(\cdot)}[f] \in C^1(\bar{\Omega}_n; \mathbb{R})$, then*

$$\int_{\Omega_n} f(t) \cdot {}^C_{a_i} D_{t_i}^{\alpha_i(\cdot)}[\eta](t)\, dt$$

$$= \int_{\partial\Omega_n} \eta(t) \cdot {}_{t_i} I_{b_i}^{1-\alpha_i(\cdot)}[f](t) \cdot \nu^i\, d(\partial\Omega_n) + \int_{\Omega_n} \eta(t) \cdot {}_{t_i} D_{b_i}^{\alpha_i(\cdot)}[f](t)\, dt.$$

4.7.2 Fundamental Problem

In this subsection, we use the notion of generalized fractional gradient.

Definition 4.49 Let $n \in \mathbb{N}$ and $P = (P_1, \ldots, P_n)$, $P_i = \langle a_i, t, b_i, \lambda_i, \mu_i \rangle$. We define the generalized fractional gradient of a function $f : \mathbb{R}^n \to \mathbb{R}$ with respect to the generalized fractional operator T by

$$\nabla_{T_P}[f] := \sum_{i=1}^n e_i \cdot T_{P_i}[f],$$

where $\{e_i : i = 1, \ldots, n\}$ denotes the standard basis in $\mathbb{R}^n$.

For $n \in \mathbb{N}$ let us assume that $P_i = \langle a_i, t_i, b_i, \lambda_i, \mu_i \rangle$ and $P = (P_1, \ldots, P_n)$, $y : \mathbb{R}^n \to \mathbb{R}$, and $\zeta : \partial\Omega_n \to \mathbb{R}$ is a given function. Consider the following functional:

$$
\begin{aligned}
\mathcal{I} : \mathcal{A}(\zeta) &\longrightarrow \mathbb{R} \\
y &\longmapsto \int_{\Omega_n} F(y(t), \nabla_{K_P}[y](t), \nabla[y](t), \nabla_{B_P}[y](t), t)\, dt
\end{aligned}
\tag{4.55}
$$

where

$$\mathcal{A}(\zeta) := \left\{ y \in C^1(\bar\Omega_n; \mathbb{R}) : y|_{\partial\Omega_n} = \zeta, \ K_{P_i}[y], B_{P_i}[y] \in C(\bar\Omega_n; \mathbb{R}), i = 1, \ldots, n \right\},$$

∇ denotes the classical gradient operator, ∇_{K_P} and ∇_{B_P} are generalized fractional gradient operators such that K_{P_i} is the generalized partial fractional integral with the kernel $k_i = k_i(t_i - \tau)$, $k_i \in L^1(0, b_i - a_i; \mathbb{R})$, and B_{P_i} is the generalized partial fractional derivative of Caputo type satisfying $B_{P_i} = K_{P_i} \circ \frac{\partial}{\partial t_i}$, for $i = 1, \ldots, n$. Moreover, we assume that F is a Lagrangian of class C^1:

$$
\begin{aligned}
F : \ \mathbb{R} \times \mathbb{R}^{3n} \times \bar\Omega_n &\longrightarrow \mathbb{R} \\
(x_1, x_2, x_3, x_4, t) &\longmapsto F(x_1, x_2, x_3, x_4, t),
\end{aligned}
$$

and

- $K_{P_i^*}\left[\partial_{1+i} F(y(\tau), \nabla_{K_P}[y](\tau), \nabla[y](\tau), \nabla_{B_P}[y](\tau), \tau)\right] \in C(\bar{\Omega}_n; \mathbb{R})$,
- $t \mapsto \partial_{1+n+i} F(y(t), \nabla_{K_P}[y](t), \nabla[y](t), \nabla_{B_P}[y](t), t) \in C^1(\bar{\Omega}_n; \mathbb{R})$,
- $K_{P_i^*}\left[\partial_{1+2n+i} F(y(\tau), \nabla_{K_P}[y](\tau), \nabla[y](\tau), \nabla_{B_P}[y](\tau), \tau)\right] \in C^1(\bar{\Omega}_n; \mathbb{R})$,

where $i = 1, \ldots, n$.

The following theorem states that if a function minimizes functional (4.55), then it necessarily must satisfy (4.56). This means that Eq. (4.56) determines candidates to solve the problem of minimizing functional (4.55).

Theorem 4.50 *Suppose that $\bar{y} \in \mathcal{A}(\zeta)$ is a minimizer of (4.55). Then, $\bar{y}$ satisfies the following generalized Euler–Lagrange equation:*

$$\partial_1 F(\star_y)(t) + \sum_{i=1}^{n} \left(K_{P_i^*}[\partial_{1+i} F(\star_y)(\tau)](t) - \frac{\partial}{\partial t_i}\left(\partial_{1+n+i} F(\star_y)(t)\right) \right.$$

$$\left. - A_{P_i^*}[\partial_{1+2n+i} F(\star_y)(\tau)](t) \right) = 0, \quad t \in \Omega_n, \qquad (4.56)$$

where $(\star_y)(t) = (y(t), \nabla_{K_P}[y](t), \nabla[y](t), \nabla_{B_P}[y](t), t)$.

Proof Let $\bar{y} \in \mathcal{A}(\zeta)$ be a minimizer of (4.55). Then, for any $|h| \leq \varepsilon$ and every $\eta \in \mathcal{A}(0)$, it satisfies

$$\mathcal{I}(\bar{y}) \leq \mathcal{I}(\bar{y} + h\eta).$$

Now, let us define the following function:

$$\phi_{\bar{y},\eta} : [-\varepsilon, \varepsilon] \longrightarrow \mathbb{R}$$
$$h \longmapsto \mathcal{I}(\bar{y} + h\eta),$$

where

$$\mathcal{I}(\bar{y}+h\eta) = \int_{\Omega_n} F(\bar{y}(t)+h\eta(t), \nabla_{K_P}[\bar{y}+h\eta](t), \nabla[\bar{y}+h\eta](t), \nabla_{B_P}[\bar{y}+h\eta](t), t)\, dt.$$

Because $\phi_{\bar{y},\eta} \in C^1([-\varepsilon, \varepsilon]; \mathbb{R})$ and

$$\phi_{\bar{y},\eta}(0) \leq \phi_{\bar{y},\eta}(h), \quad |h| \leq \varepsilon,$$

one has

$$\phi'_{\bar{y},\eta}(0) = \frac{d}{dh}\mathcal{I}(\bar{y} + h\eta)\bigg|_{h=0} = 0.$$

Moreover, using the chain rule, we obtain

$$\int_{\Omega_n} \partial_1 F(\star_{\bar{y}})(t) \cdot \eta(t) + \sum_{i=1}^{n} \left(\partial_{1+i} F(\star_{\bar{y}})(t) \cdot K_{P_i}[\eta](t) + \partial_{1+n+i} F(\star_{\bar{y}})(t) \cdot \frac{\partial \eta(t)}{\partial t_i} \right.$$

$$\left. + \partial_{1+2n+i} F(\star_{\bar{y}})(t) \cdot B_{P_i}[\eta](t) \right) dt = 0.$$

Finally, Theorem 4.43 implies that

$$\int_{\Omega_n} \left(\partial_1 F(\star_{\bar{y}})(t) + \sum_{i=1}^{n} K_{P_i^*} \left[\partial_{1+i} F(\star_{\bar{y}})(\tau) \right](t) \right) \cdot \eta(t)$$

$$+ \sum_{i=1}^{n} \left(\partial_{1+n+i} F(\star_{\bar{y}})(t) + K_{P_i^*} \left[\partial_{1+2n+i} F(\star_{\bar{y}})(\tau) \right](t) \right) \cdot \frac{\partial \eta(t)}{\partial t_i} dt = 0$$

and by the classical integration by parts formula (see, e.g., (Evans 2010)) and the fundamental lemma of the calculus of variations (see, e.g., Theorem 1.24 of (Dacorogna 2004)), we arrive to the validity of (4.56) along $\bar{y}$.

Example 4.51 Consider a motion of medium whose displacement is described as a scalar function $y(t, x)$, where $x = (x^1, x^2)$. For example, this function may represent the transverse displacement of a membrane. Suppose that the kinetic energy T and the potential energy V of the medium are given by

$$T \left(\frac{\partial y(t, x)}{\partial t} \right) = \frac{1}{2} \int \rho(x) \left(\frac{\partial y(t, x)}{\partial t} \right)^2 dx,$$

$$V(y) = \frac{1}{2} \int k(x) \| \nabla[y](t, x) \|^2 dx,$$

where $\rho(x)$ is the mass density and $k(x)$ is the stiffness, both assumed positive. Then, the classical action functional is

$$\mathcal{I}(y) = \frac{1}{2} \int_{\Omega} \left(\rho(x) \left(\frac{\partial y(t, x)}{\partial t} \right)^2 - k(x) \| \nabla[y](t, x) \|^2 \right) dx dt.$$

We shall illustrate what are the Euler–Lagrange equations when the Lagrangian density depends on generalized fractional derivatives. When we have the Lagrangian with the kinetic term depending on the operator B_{P_1}, with $P_1 = \langle a_1, t, b_1, \lambda, \mu \rangle$, then the fractional action functional has the form

$$\mathcal{I}(y) = \frac{1}{2} \int_{\Omega_3} \left[\rho(x) \left(B_{P_1}[y](t, x) \right)^2 - k(x) \| \nabla[y](t, x) \|^2 \right] dx dt. \tag{4.57}$$

The generalized fractional Euler–Lagrange equation satisfied by an extremal of (4.57) is $-\rho(x)A_{P_1^*}\left[B_{P_1}[y](\tau, s)\right](t, x) - \nabla\left[k(s)\nabla[y](\tau, s)\right](t, x) = 0$. If ρ and k are constants, then equation $A_{P_1^*}\left[B_{P_1}[y](\tau, s)\right](t, x) - c^2\Delta[y](t, x) = 0$, where $c^2 = k/\rho$, can be called the generalized time-fractional wave equation. Now assume that the kinetic energy and the potential energy depend on B_{P_1} and ∇_{B_P} operators, respectively, where $P = (P_2, P_3)$. Then, the action functional for the system has the form

$$\mathcal{I}(y) = \frac{1}{2}\int\limits_{\Omega_3}\left[\rho\left(B_{P_1}[y](t, x)\right)^2 - k\left\|\nabla_{B_P}[y](t, x)\right\|^2\right]dxdt. \tag{4.58}$$

The generalized fractional Euler–Lagrange equation of (4.58) is

$$-\rho A_{P_1^*}\left[B_{P_1}[y](\tau, s)\right](t, x) + \sum_{i=2}^{3}A_{P_i^*}\left[k(s)B_{P_i}[y](\tau, s)\right](t, x) = 0.$$

If ρ and k are constants, then

$$A_{P_1^*}\left[B_{P_1}[y](\tau, s)\right](t, x) - c^2\left(\sum_{i=2}^{3}A_{P_i^*}\left[B_{P_i}[y](\tau, s)\right](t, x)\right) = 0$$

can be called the generalized space- and time-fractional wave equation.

Corollary 4.52 *Let $\alpha = (\alpha_1, \ldots, \alpha_n) \in (0, 1)^n$ and let $\bar{y} \in C^1(\bar{\Omega}_n; \mathbb{R})$ be a minimizer of the functional*

$$\mathcal{I}(y) = \int\limits_{\Omega_n}F(y(t), \nabla_{I^{1-\alpha}}[y](t), \nabla[y](t), \nabla_{D^\alpha}[y](t), t)\,dt \tag{4.59}$$

satisfying

$$y(t)|_{\partial\Omega_n} = \zeta(t), \tag{4.60}$$

where $\zeta : \partial\Omega_n \to \mathbb{R}$ is a given function,

$$\nabla_{I^{1-\alpha}} = \sum_{i=1}^{n}e_i \cdot {}_{a_i}I_{t_i}^{1-\alpha_i}, \quad \nabla_{D^\alpha} = \sum_{i=1}^{n}e_i \cdot {}_{a_i}^{C}D_{t_i}^{\alpha_i},$$

F is of class C^1 and

- *${}_t I_{b_i}^{1-\alpha_i}\left[\partial_{1+i}F(y(\tau), \nabla_{I^{1-\alpha}}[y](\tau), \nabla[y](\tau), \nabla_{D^\alpha}[y](\tau), \tau)\right]$ is continuous on $\bar{\Omega}_n$,*
- *$t \mapsto \partial_{1+n+i}F(y(t), \nabla_{I^{1-\alpha}}[y](t), \nabla[y](t), \nabla_{D^\alpha}[y](t), t)$ is continuously differentiable on $\bar{\Omega}_n$,*

- $_t I_{b_i}^{1-\alpha_i} \left[\partial_{1+2n+i} F(y(\tau), \nabla_{I^{1-\alpha}}[y](\tau), \nabla[y](\tau), \nabla_{D^\alpha}[y](\tau), \tau) \right]$ is continuously differentiable on $\bar{\Omega}_n$.

Then, $\bar{y}$ satisfies the following fractional Euler–Lagrange equation, $t \in \Omega_n$:

$$\partial_1 F(y(t), \nabla_{I^{1-\alpha}}[y](t), \nabla[y](t), \nabla_{D^\alpha}[y](t), t)$$

$$+ \sum_{i=1}^{n} \left({}_t I_{b_i}^{1-\alpha_i} \left[\partial_{1+i} F(y(\tau), \nabla_{I^{1-\alpha}}[y](\tau), \nabla[y](\tau), \nabla_{D^\alpha}[y](\tau), \tau) \right] (t) \right.$$

$$- \frac{\partial}{\partial t_i} \left(\partial_{1+n+i} F(y(t), \nabla_{I^{1-\alpha}}[y](t), \nabla[y](t), \nabla_{D^\alpha}[y](t), t) \right)$$

$$\left. + {}_t D_{b_i}^{\alpha_i} \left[\partial_{1+2n+i} F(y(\tau), \nabla_{I^{1-\alpha}}[y](\tau), \nabla[y](\tau), \nabla_{D^\alpha}[y](\tau), \tau) \right] (t) \right) = 0.$$

$$(4.61)$$

Corollary 4.53 For $i = 1, \ldots, n$, suppose that $\alpha_i : [0, b_i - a_i] \to [0, 1]$. Let

$$\nabla_I = \sum_{i=1}^{n} e_i \cdot {}_{a_i} I_{t_i}^{1-\alpha_i(\cdot)}, \quad \nabla_D = \sum_{i=1}^{n} e_i \cdot {}_{a_i}^{C} D_{t_i}^{\alpha_i(\cdot)}.$$

If $\bar{y} \in C^1(\bar{\Omega}_n; \mathbb{R})$ minimizes the functional

$$\mathcal{I}(y) = \int_{\Omega_n} F(y(t), \nabla_I[y](t), \nabla[y](t), \nabla_D[y](t), t) \, dt \qquad (4.62)$$

subject to the boundary condition $y(t)|_{\partial\Omega_n} = \zeta(t)$, where $\zeta : \partial\Omega_n \to \mathbb{R}$ is a given function, ${}_{a_i} I_{t_i}^{1-\alpha_i(\cdot)}[y], {}_{a_i}^{C} D_{t_i}^{\alpha_i(\cdot)} \in C(\bar{\Omega}_n; \mathbb{R})$, F is of class C^1 and

- ${}_{t_i} I_{b_i}^{1-\alpha_i(\cdot)} \left[\partial_{1+i} F(y(\tau), \nabla_I[y](\tau), \nabla[y](\tau), \nabla_D[y](\tau), \tau) \right]$ is continuous on $\bar{\Omega}_n$,
- $t \mapsto \partial_{1+n+i} F(y(t), \nabla_I[y](t), \nabla[y](t), \nabla_D[y](t), t)$ is continuously differentiable on $\bar{\Omega}_n$,
- ${}_{t_i} I_{b_i}^{1-\alpha_i(\cdot)} \left[\partial_{1+2n+i} F(y(\tau), \nabla_I[y](\tau), \nabla[y](\tau), \nabla_D[y](\tau), \tau) \right]$ is continuously differentiable on $\bar{\Omega}_n$,

then $\bar{y}$ satisfies the following equation in $t \in \Omega_n$:

$$\partial_1 F(y(t), \nabla_I[y](t), \nabla[y](t), \nabla_D[y](t), t)$$

$$+ \sum_{i=1}^{n} \left({}_{t_i} I_{b_i}^{1-\alpha_i(\cdot)} \left[\partial_{1+i} F(y(\tau), \nabla_I[y](\tau), \nabla[y](\tau), \nabla_D[y](\tau), \tau) \right] (t) \right.$$

$$- \frac{\partial}{\partial t_i} \left(\partial_{1+n+i} F(y(t), \nabla_I[y](t), \nabla[y](t), \nabla_D[y](t), t) \right)$$

$$\left. + {}_{t_i} D_{b_i}^{\alpha_i(\cdot)} \left[\partial_{1+2n+i} F(y(\tau), \nabla_I[y](\tau), \nabla[y](\tau), \nabla_D[y](\tau), \tau) \right] (t) \right) = 0.$$

$$(4.63)$$

Theorem 4.54 *Suppose that* $\bar{y} \in \mathcal{A}(\zeta)$ *satisfies* (4.56) *and function*

$$(x_1, x_2, x_3, x_4) \rightarrow F(x_1, x_2, x_3, x_4, t)$$

is convex for every $t \in \bar{\Omega}_n$. *Then,* $\bar{y}$ *is a minimizer of functional* (4.55).

Proof Let $\bar{y} \in \mathcal{A}(\zeta)$ be a function satisfying Eq. (4.56) and such that $(x_1, x_2, x_3, x_4) \rightarrow F(x_1, x_2, x_3, x_4, t)$ is convex for every $t \in \bar{\Omega}_n$. Then, the following inequality holds:

$$
\begin{aligned}
\mathcal{I}(y) \geq \mathcal{I}(\bar{y}) & \\
+ \int_{\Omega_n} & \left(\partial_1 F (y - \bar{y}) + \sum_{i=1}^{n} \left[\partial_{1+i} F \cdot K_{P_i}[y - \bar{y}] \right. \right. \\
& \left. \left. + \partial_{1+n+i} F \frac{\partial}{\partial t_i}[y - \bar{y}] + \partial_{1+2n+i} F \cdot B_{P_i}[y - \bar{y}] \right] \right) dt,
\end{aligned}
$$

where functions $\partial_i F$ are evaluated at $(\bar{y}, \nabla_{K_P}[\bar{y}], \nabla[\bar{y}], \nabla_{B_P}[\bar{y}], t)$, for $i = 1, \ldots, 3n + 1$. Moreover, using the classical integration by parts formula, as well as Theorem 4.43 and the fact that $y - \bar{y}|_{\partial \Omega_n} = 0$, we obtain

$$
\begin{aligned}
\mathcal{I}(y) \geq \mathcal{I}(\bar{y}) & \\
+ \int_{\Omega_n} & \left(\partial_1 F + \sum_{i=1}^{n} \left[K_{P_i^*} \left[\partial_{1+i} F \right] + \frac{\partial}{\partial t_i} \left(\partial_{1+n+i} F \right) + A_{P_i^*} \left[\partial_{1+2n+i} F \right] \right] \right) (y - \bar{y}) \, dt.
\end{aligned}
$$

Finally, applying Eq. (4.56), we have $\mathcal{I}(y) \geq \mathcal{I}(\bar{y})$ for any $y \in \mathcal{A}(\zeta)$ and the proof is complete.

4.7.3 Dirichlet's Principle

One of the most important variational principles for PDEs is Dirichlet's principle for the Laplace equation. We shall present its generalized fractional counterpart. We show that the solution of the generalized fractional boundary value problem

$$
\begin{cases}
\sum_{i=1}^{n} A_{P_i^*} \left[B_{P_i}[y] \right] = 0 & \text{in } \Omega_n, \\
y = \zeta & \text{on } \partial \Omega_n,
\end{cases}
\tag{4.64}
$$

can be characterized as a minimizer of the following variational functional:

$$
\mathcal{I}(y) = \int_{\Omega_n} \left\| \nabla_{B_P}[y] \right\|^2 dt
\tag{4.65}
$$

on the set $\mathcal{A}(\zeta)$, where $\nabla_{BP} = \sum_{i=1}^{n} e_i \cdot B_{P_i}$ is the generalized fractional gradient operator such that the partial derivatives B_{P_i} have kernels $k_i = k_i(t_i - \tau)$, $k_i \in L^1(0, b_i - a_i; \mathbb{R})$, and parameter sets are given by $P_i = \langle a_i, t_i, b_i, \lambda_i, \mu_i \rangle$, $i = 1, \ldots, n$.

Remark 4.55 In the following we assume that both problems, (4.64) and minimization of (4.65) on the set $\mathcal{A}(\zeta)$, have solutions.

Theorem 4.56 (Generalized Fractional Dirichlet's Principle) *Suppose that $\bar{y} \in \mathcal{A}(\zeta)$. Then $\bar{y}$ solves the boundary value problem (4.64) if and only if $\bar{y}$ satisfies*

$$\mathcal{I}(\bar{y}) = \min_{y \in \mathcal{A}(\zeta)} \mathcal{I}(y). \tag{4.66}$$

Proof Theorem 4.56 is a simple consequence of Theorems 4.50 and 4.54.

Theorem 4.57 *There exists at most one solution $\bar{y} \in \mathcal{A}(\zeta)$ to problem (4.64).*

Proof Let $\bar{y} \in \mathcal{A}(\zeta)$ be a solution to problem (4.64). Assume that $\hat{y}$ is another solution to problem (4.64). Then, $w = \bar{y} - \hat{y} \neq 0$ and

$$0 = -\int_{\Omega_n} w \cdot \sum_{i=1}^{n} A_{P_i^*} \left[B_{P_i}[w] \right] dt.$$

By the classical integration by parts formula and Theorem 4.43, and since $w|_{\partial\Omega_n} = 0$, we have

$$0 = \int_{\Omega_n} w \cdot \sum_{i=1}^{n} \left(B_{P_i}[w] \right)^2 dt = \int_{\Omega_n} w \cdot \left\| \nabla_{BP}[w] \right\|^2 dt.$$

Note that $\left\| \nabla_{BP}[w] \right\|^2$ is a positive definite quantity. The volume integral of a positive definite quantity is equal to zero only in the case when this quantity is zero itself throughout the volume. Thus $\nabla_{BP}[w] = 0$. Since w is twice continuously differentiable and $k_i \in L^1(0, b_i - a_i; \mathbb{R})$ we have $\frac{\partial}{\partial t_i} w(t) = 0$, $i = 1, \ldots, n$, i.e., $\nabla[w] = 0$. Because $w = 0$ on $\partial\Omega_n$, we deduce that $w = 0$. In other words, $\bar{y} = \hat{y}$.

4.7.4 Isoperimetric Problem

Suppose that $y : \mathbb{R}^n \to \mathbb{R}$, $P = \langle a_i, t_i, b_i, \lambda_i, \mu_i \rangle$, $P = (P_1, \ldots, P_n)$, and $\zeta : \partial\Omega_n \to \mathbb{R}$ is a given curve. Let us define the following functional:

$$\mathcal{J} : \mathcal{A}(\zeta) \longrightarrow \mathbb{R} \tag{4.67}$$
$$y \longmapsto \int_{\Omega_n} G(y(t), \nabla_{KP}[y](t), \nabla[y](t), \nabla_{BP}[y](t), t) \, dt,$$

where operators $\nabla_{KP}, \nabla, \nabla_{BP}$, are defined as in Sec. 4.7.2, and function G has the same properties as in the case of functional (4.55). The next theorem gives a necessary optimality condition for a function to be a minimizer of functional (4.55) subject to the isoperimetric constraint $\mathcal{J}(y) = \xi$.

Theorem 4.58 *Let us assume that $\bar{y}$ minimizes functional (4.55) on the set*

$$\mathcal{A}_\xi(\zeta) := \{y \in \mathcal{A}(\zeta) : \mathcal{J}(y) = \xi\}.$$

Then, one can find a real constant λ_0 such that, for $H = F - \lambda_0 G$, equality

$$\partial_1 H(\star_{\bar{y}})(t) + \sum_{i=1}^n \bigg(K_{P_i^*}[\partial_{1+i} H(\star_{\bar{y}})(\tau)](t)$$

$$- \frac{\partial}{\partial t_i} \left(\partial_{1+n+i} H(\star_{\bar{y}})(t)\right) - A_{P_i^*}[\partial_{1+2n+i} H(\star_{\bar{y}})(\tau)](t)\bigg) = 0$$

(4.68)

holds, provided that

$$\partial_1 G(\star_{\bar{y}})(t) + \sum_{i=1}^n \bigg(K_{P_i^*}[\partial_{1+i} G(\star_{\bar{y}})(\tau)](t)$$

$$- \frac{\partial}{\partial t_i} \left(\partial_{1+n+i} G(\star_{\bar{y}})(t)\right) - A_{P_i^*}[\partial_{1+2n+i} G(\star_{\bar{y}})(\tau)](t)\bigg) \neq 0,$$

(4.69)

where $(\star_{\bar{y}})(t) = (\bar{y}(t), \nabla_{KP}[\bar{y}](t), \nabla[\bar{y}](t), \nabla_{BP}[\bar{y}](t), t)$.

Proof The fundamental lemma of the calculus of variations, and hypothesis (4.69), imply that there exists $\eta_2 \in \mathcal{A}(0)$ so that

$$\int_{\Omega_n} \left(\partial_1 G(\star_{\bar{y}})(t) + \sum_{i=1}^n K_{P_i^*}\left[\partial_{1+i} G(\star_{\bar{y}})(\tau)\right](t)\right) \cdot \eta_2(t)$$

$$+ \sum_{i=1}^n \left(\partial_{1+n+i} G(\star_{\bar{y}})(t) + K_{P_i^*}\left[\partial_{1+2n+i} G(\star_{\bar{y}})(\tau)\right](t)\right) \cdot \frac{\partial \eta_2(t)}{\partial t_i} \, dt = 1.$$

Now, with function η_2 and an arbitrary $\eta_1 \in \mathcal{A}(0)$, let us define

$$\phi : [-\varepsilon_1, \varepsilon_1] \times [-\varepsilon_2, \varepsilon_2] \longrightarrow \mathbb{R}$$
$$(h_1, h_2) \longmapsto \mathcal{I}(\bar{y} + h_1\eta_1 + h_2\eta_2)$$

and

$$\psi : [-\varepsilon_1, \varepsilon_1] \times [-\varepsilon_2, \varepsilon_2] \longrightarrow \mathbb{R}$$
$$(h_1, h_2) \longmapsto \mathcal{J}(\bar{y} + h_1\eta_1 + h_2\eta_2) - \xi .$$

Note that $\psi(0, 0) = 0$ and that

$$\frac{\partial\psi}{\partial h_2}\bigg|_{(0,0)} = \int_{\Omega_n} \left(\partial_1 G(\star_{\bar{y}})(t) + \sum_{i=1}^{n} K_{P_i^*}\left[\partial_{1+i} G(\star_{\bar{y}})(\tau)\right](t) \right) \cdot \eta_2(t)$$

$$+ \sum_{i=1}^{n} \left(\partial_{1+n+i} G(\star_{\bar{y}})(t) + K_{P_i^*}\left[\partial_{1+2n+i} G(\star_{\bar{y}})(\tau)\right](t) \right) \cdot \frac{\partial\eta_2(t)}{\partial t_i}\, dt = 1.$$

The implicit function theorem implies that there is $\epsilon_0 > 0$ and a function $s \in C^1([-\varepsilon_0, \varepsilon_0]; \mathbb{R})$ with $s(0) = 0$ such that $\psi(h_1, s(h_1)) = 0$, $|h_1| \le \varepsilon_0$, and then $\bar{y} + h_1\eta_1 + s(h_1)\eta_2 \in \mathcal{A}_\xi(\zeta)$. Moreover,

$$\frac{\partial\psi}{\partial h_1} + \frac{\partial\psi}{\partial h_2} \cdot s'(h_1) = 0, \quad |h_1| \le \varepsilon_0,$$

and then

$$s'(0) = -\frac{\partial\psi}{\partial h_1}\bigg|_{(0,0)}.$$

Because $\bar{y} \in \mathcal{A}(\zeta)$ is a minimizer of $\mathcal{I}$ we have $\phi(0, 0) \le \phi(h_1, s(h_1))$, $|h_1| \le \varepsilon_0$, and hence

$$\frac{\partial\phi}{\partial h_1}\bigg|_{(0,0)} + \frac{\partial\phi}{\partial h_2}\bigg|_{(0,0)} \cdot s'(0) = 0.$$

Letting $\lambda_0 = \frac{\partial\phi}{\partial h_2}\big|_{(0,0)}$ be the Lagrange multiplier we find

$$\frac{\partial\phi}{\partial h_1}\bigg|_{(0,0)} - \lambda_0 \frac{\partial\psi}{\partial h_1}\bigg|_{(0,0)} = 0$$

or, in other words,

$$\int_{\Omega_n} \left[\left(\partial_1 F(\star_{\bar{y}})(t) + \sum_{i=1}^{n} K_{P_i^*}\left[\partial_{1+i} F(\star_{\bar{y}})(\tau)\right](t) \right) \cdot \eta_1(t) \right.$$

$$+ \sum_{i=1}^{n} \left(\partial_{1+n+i} F(\star_{\bar{y}})(t) + K_{P_i^*}\left[\partial_{1+2n+i} F(\star_{\bar{y}})(\tau)\right](t) \right) \cdot \frac{\partial\eta_1(t)}{\partial t_i} \right]$$

$$- \lambda_0 \left[\left(\partial_1 G(\star_{\bar{y}})(t) + \sum_{i=1}^{n} K_{P_i^*}\left[\partial_{1+i} G(\star_{\bar{y}})(\tau)\right](t) \right) \cdot \eta_1(t) \right.$$

$$+ \sum_{i=1}^{n} \left(\partial_{1+n+i} G(\star_{\bar{y}})(t) + K_{P_i^*} \left[\partial_{1+2n+i} G(\star_{\bar{y}})(\tau) \right](t) \right) \cdot \frac{\partial \eta_1(t)}{\partial t_i} \right] dt = 0.$$

Finally, applying the integration by parts formula for classical and fractional derivatives, and by the fundamental lemma of the calculus of variations, we obtain (4.68).

Corollary 4.59 *Let us assume that* $\alpha = (\alpha_1, \dots, \alpha_n) \in (0,1)^n$ *and* $\bar{y} \in C^1(\bar{\Omega}_n; \mathbb{R})$ *is minimizer of functional* (4.59) *subject to the isoperimetric constraint* $\mathcal{J}(y) = \xi$, *where*

$$\mathcal{J}(y) = \int_{\Omega_n} G(y(t), \nabla_{I^{1-\alpha}}[y](t), \nabla[y](t), \nabla_{D^\alpha}[y](t), t) \, dt \qquad (4.70)$$

and boundary condition (4.60). *Moreover,*

- G *is of class* C^1,
- $_{t_i} I_{b_i}^{1-\alpha_i} \left[\partial_{1+i} G(y(\tau), \nabla_I[y](\tau), \nabla[y](\tau), \nabla_D[y](\tau), \tau) \right]$ *is continuous on* $\bar{\Omega}_n$,
- $t \mapsto \partial_{1+n+i} G(y(t), \nabla_I[y](t), \nabla[y](t), \nabla_D[y](t), t)$ *is continuously differentiable on* $\bar{\Omega}_n$,
- $_{t_i} I_{b_i}^{1-\alpha_i} \left[\partial_{1+2n+i} G(y(\tau), \nabla_I[y](\tau), \nabla[y](\tau), \nabla_D[y](\tau), \tau) \right]$ *is continuously differentiable on* $\bar{\Omega}_n$.

Then, if $\bar{y}$ *is not an extremal for functional* (4.70), *we can find* $\lambda_0 \in \mathbb{R}$ *such that*

$$\partial_1 H(\bar{y}(t), \nabla_{I^{1-\alpha}}[\bar{y}](t), \nabla[\bar{y}](t), \nabla_{D^\alpha}[\bar{y}](t), t)$$

$$+ \sum_{i=1}^{n} \Big({_{t_i}} I_{b_i}^{1-\alpha_i} \left[\partial_{1+i} H(\bar{y}(\tau), \nabla_{I^{1-\alpha}}[\bar{y}](\tau), \nabla[\bar{y}](\tau), \nabla_{D^\alpha}[\bar{y}](\tau), \tau) \right](t)$$

$$- \frac{\partial}{\partial t_i} \left(\partial_{1+n+i} H(\bar{y}(t), \nabla_{I^{1-\alpha}}[\bar{y}](t), \nabla[\bar{y}](t), \nabla_{D^\alpha}[\bar{y}](t), t) \right)$$

$$+ {_{t_i}} D_{b_i}^{\alpha_i} \left[\partial_{1+2n+i} H(\bar{y}(\tau), \nabla_{I^{1-\alpha}}[\bar{y}](\tau), \nabla[\bar{y}](\tau), \nabla_{D^\alpha}[\bar{y}](\tau), \tau) \right](t) \Big) = 0,$$

$$t \in \Omega_n,$$

is satisfied, where $H = F - \lambda_0 G$.

Corollary 4.60 *Suppose that* $\alpha_i : [0, b_i - a_i] \to [0,1]$. *If* $\bar{y} \in C^1(\bar{\Omega}_n; \mathbb{R})$ *minimizes* (4.62) *subject to* (4.60) *and*

$$\mathcal{J}(y) = \int_{\Omega_n} G(y(t), \nabla_I[y](t), \nabla[y](t), \nabla_D[y](t), t) \, dt = \xi,$$

where $_{a_i} I_{t_i}^{1-\alpha_i(\cdot)}[y]$, $_{a_i}^C D_{t_i}^{\alpha_i(\cdot)}[y] \in C(\bar{\Omega}_n; \mathbb{R})$, *and*

- G *is of class* C^1,

- $_{t_i}I_{b_i}^{1-\alpha_i(\cdot)}\left[\partial_{1+i}G(y(\tau), \nabla_I[y](\tau), \nabla[y](\tau), \nabla_D[y](\tau), \tau)\right]$ is continuous on $\bar{\Omega}_n$,
- $t \mapsto \partial_{1+n+i}G(y(t), \nabla_I[y](t), \nabla[y](t), \nabla_D[y](t), t)$ is continuously differentiable on $\bar{\Omega}_n$,
- $_{t_i}I_{b_i}^{1-\alpha_i(\cdot)}\left[\partial_{1+2n+i}G(y(\tau), \nabla_I[y](\tau), \nabla[y](\tau), \nabla_D[y](\tau), \tau)\right]$ is continuously differentiable on $\bar{\Omega}_n$,

for $i = 1, \ldots, n$, then there is $\lambda_0 \in \mathbb{R}$ such that, for $H = F - \lambda_0 G$, $\bar{y}$ satisfies the following equality:

$$\partial_1 H(\bar{y}(t), \nabla_I[\bar{y}](t), \nabla[\bar{y}](t), \nabla_D[\bar{y}](t), t)$$

$$+ \sum_{i=1}^{n}\left({}_{t_i}I_{b_i}^{1-\alpha_i(\cdot)}\left[\partial_{1+i}H(\bar{y}(\tau), \nabla_I[\bar{y}](\tau), \nabla[\bar{y}](\tau), \nabla_D[\bar{y}](\tau), \tau)\right](t)\right.$$

$$- \frac{\partial}{\partial t_i}(\partial_{1+n+i}H(\bar{y}(t), \nabla_I[\bar{y}](t), \nabla[\bar{y}](t), \nabla_D[\bar{y}](t), t))$$

$$\left. + {}_{t_i}D_{b_i}^{\alpha_i(\cdot)}\left[\partial_{1+2n+i}H(\bar{y}(\tau), \nabla_I[\bar{y}](\tau), \nabla[\bar{y}](\tau), \nabla_D[\bar{y}](\tau), \tau)\right](t)\right) = 0,$$

$$t \in \Omega_n,$$

provided that $\bar{y}$ is not a solution to the Euler–Lagrange equation associated to $\mathcal{J}$.

4.7.5 Noether's Theorem

In Sect. 4.5 of this book, we have proved a generalized fractional version of Noether's theorem. That is, assuming invariance of Lagrangian under changes in the coordinate system, we showed that its extremal must satisfy Eq. (4.43). In this section, we prove a generalized multidimensional fractional Noether's theorem. As before, we start with definitions of extremal and invariance.

Definition 4.61 A function $y \in C^1(\bar{\Omega}_n; \mathbb{R})$ such that $K_{P_i}[y], B_{P_i}[y] \in C(\bar{\Omega}_n; \mathbb{R})$, $i = 1, \ldots, n$ satisfying Eq. (4.56) is said to be a generalized fractional extremal.

We consider a one-parameter family of transformations of the form $\hat{y} = \phi(\theta, t, y)$, where ϕ is a map of class C^2:

$$\phi : [-\varepsilon, \varepsilon] \times \bar{\Omega}_n \times \mathbb{R} \longrightarrow \mathbb{R}$$
$$(\theta, t, y) \longmapsto \phi(\theta, t, y)$$

such that $\phi(0, t, y) = y$. Note that, using Taylor's expansion of $\phi(\theta, t, y)$ in a neighborhood of 0, one has $\hat{y} = \phi(0, t, y) + \theta\frac{\partial}{\partial\theta}\phi(0, t, y) + o(\theta)$, provided that $\theta \in [-\varepsilon, \varepsilon]$. Moreover, having in mind that $\phi(0, t, y) = y$, and denoting $\xi(t, y) = \frac{\partial}{\partial\theta}\phi(0, t, y)$, one has

$$\hat{y} = y + \theta\xi(t, y) + o(\theta), \tag{4.71}$$

so that the linear approximation to the transformation is $\hat{y} \approx y + \theta \xi(t, y)$ for $\theta \in [-\varepsilon, \varepsilon]$. Let $y : \bar{\Omega}_n \to \mathbb{R}$ be given by $y = y(t)$. Then, for sufficiently small θ, the transformation (4.71) carries the curve $y = y(t)$ into a family of curves $\hat{y} = \hat{y}(t) = \phi(\theta, t, y(t))$. Now, let us introduce the notion of invariance.

Definition 4.62 We say that the Lagrangian F is invariant under the one-parameter family of infinitesimal transformations (4.71), where ξ is such that $t \mapsto \xi(t, y(t)) \in C^1\left(\bar{\Omega}_n; \mathbb{R}\right)$ with $K_{P_i}\left[\tau \mapsto \xi(\tau, y(\tau))\right], B_{P_i}\left[\tau \mapsto \xi(\tau, y(\tau))\right] \in C\left(\bar{\Omega}_n; \mathbb{R}\right)$, $i = 1, \ldots, n$, if

$$F(y(t), \nabla_{K_P}[y](t), \nabla[y](t), \nabla_{B_P}[y](t), t)$$

$$= F(\hat{y}(t), \nabla_{K_P}[\hat{y}](t), \nabla[\hat{y}](t), \nabla_{B_P}[\hat{y}](t), t), \tag{4.72}$$

for all $\theta \in [-\varepsilon, \varepsilon]$, and all $y \in C^1\left(\bar{\Omega}_n; \mathbb{R}\right)$ with $K_{P_i}[y], B_{P_i}[y] \in C\left(\bar{\Omega}_n; \mathbb{R}\right)$, $i = 1, \ldots, n$.

Similarly to Sect. 4.5, we want to state Noether's theorem in a compact form. For that we introduce the following two bilinear operators:

$$\mathbf{D}_i[f, g] := f \cdot A_{P_i^*}[g] + g \cdot B_{P_i}[f], \tag{4.73}$$

$$\mathbf{I}_i[f, g] := -f \cdot K_{P_i^*}[g] + g \cdot K_{P_i}[f], \tag{4.74}$$

where $i = 1, \ldots, n$.

Now we are ready to state the generalized fractional Noether's theorem.

Theorem 4.63 (Generalized Multidimensional Fractional Noether's Theorem) *Let F be invariant under the one-parameter family of infinitesimal transformations (4.71). Then, for every generalized fractional extremal y, the following equality holds:*

$$\sum_{i=1}^{n}\left(\mathbf{I}_i\left[\xi(t, y(t)), \partial_{1+i} F(\star_y)(t)\right] + \frac{\partial}{\partial t_i}\left(\xi(t, y(t)) \cdot \partial_{1+n+i} F(\star_y)(t)\right)\right.$$

$$\left. + \mathbf{D}_i\left[\xi(t, y(t)), \partial_{1+2n+i} F(\star_y)(t)\right]\right) = 0, \quad t \in \Omega_n, \tag{4.75}$$

where $(\star_y)(t) = (y(t), \nabla_{K_P}[y](t), \nabla[y](t), \nabla_{B_P}[y](t), t)$.

Proof By Eq. (4.72) one has

$$\frac{d}{d\theta}\left[F(\hat{y}(t), \nabla_{K_P}[\hat{y}](t), \nabla[\hat{y}](t), \nabla_{B_P}[\hat{y}](t), t)\right]\bigg|_{\theta=0} = 0.$$

The usual chain rule implies

$$\partial_1 F(\star_{\hat{y}})(t) \cdot \frac{d}{d\theta} \hat{y}(t) + \sum_{i=1}^{n} \partial_{1+i} F(\star_{\hat{y}})(t) \cdot \frac{d}{d\theta} K_{P_i}[\hat{y}](t)$$

$$+ \partial_{1+n+i} F(\star_{\hat{y}})(t) \cdot \frac{d}{d\theta} \frac{\partial}{\partial t_i} \hat{y}(t) + \partial_{1+2n+i} F(\star_{\hat{y}})(t) \cdot \frac{d}{d\theta} B_{P_i}[\hat{y}](t)\bigg|_{\theta=0} = 0.$$

By linearity of $K_{P_i}, B_{P_i}, i = 1, \ldots, n$ differentiating with respect to θ, and putting $\theta = 0$, we have

$$\partial_1 F(\star_y)(t) \cdot \xi(t, y(t)) + \sum_{i=1}^{n} \partial_{1+i} F(\star_y)(t) \cdot K_{P_i}[\tau \mapsto \xi(\tau, y(\tau))](t)$$

$$+ \partial_{1+n+i} F(\star_y)(t) \cdot \frac{\partial}{\partial t_i} \xi(t, y(t))$$

$$+ \partial_{1+2n+i} F(\star_y)(t) \cdot B_{P_i}[\tau \mapsto \xi(\tau, y(\tau))](t) = 0.$$

Now, using the generalized Euler–Lagrange equation (4.56) and formulas (4.73) and (4.74), we arrive to (4.75).

Example 4.64 Let $c \in \mathbb{R}$, $P = (P_1, \ldots, P_n)$ with $P_i = \langle a_i, t_i, b_i, \lambda_i, \mu_i \rangle$ and $y \in C^1(\bar{\Omega}_n; \mathbb{R})$ with $B_{P_i}[y] \in C(\bar{\Omega}_n; \mathbb{R})$, $i = 1, \ldots, n$. We consider a one-parameter family of infinitesimal transformations

$$\hat{y}(t) = y(t) + \varepsilon c + o(\varepsilon), \tag{4.76}$$

and the Lagrangian $F\left(\nabla_{B_P}[y](t), t\right)$. Then, $F\left(\nabla_{B_P}[y](t), t\right) = F\left(\nabla_{B_P}[\hat{y}](t), t\right)$. Hence, F is invariant under (4.76) and Theorem 4.63 asserts that

$$\sum_{i=1}^{n} \mathbf{D}_i \left[c, \partial_{n+i} F\left(\nabla_{B_P}[y](t), t\right) \right] = 0. \tag{4.77}$$

Similarly to previous sections, one can obtain from Theorem 4.63 results regarding to constant and variable order fractional integrals and derivatives.

Corollary 4.65 *Suppose that* $\alpha = (\alpha_1, \ldots, \alpha_n) \in (0, 1)^n$ *and that*

$$F(y(t), \nabla_{I^{1-\alpha}}[y](t), \nabla[y](t), \nabla_{D^\alpha}[y](t), t)$$
$$= F(\hat{y}(t), \nabla_{I^{1-\alpha}}[\hat{y}](t), \nabla[\hat{y}](t), \nabla_{D^\alpha}[\hat{y}](t), t),$$

where $y \in C^1(\bar{\Omega}_n; \mathbb{R})$ *and* $\hat{y}$ *is the family* (4.71). *Then all solutions of the Euler–Lagrange equation* (4.61) *satisfy*

$$\sum_{i=1}^{n} \left(\mathbf{I}_i^{1-\alpha_i} \left[\xi, \partial_{1+i} F \right] + \frac{\partial}{\partial t_i} (\xi \cdot \partial_{1+n+i} F) + \mathbf{D}_i^{\alpha_i} \left[\xi, \partial_{1+2n+i} F \right] \right) = 0,$$

where

$$\mathbf{D}_i^{\alpha_i}[f, g] := - f \cdot {}_{t_i}D_{b_i}^{\alpha_i}[g] + g \cdot {}_{a_i}^C D_{t_i}^{\alpha_i}[f],$$

$$\mathbf{I}_i^{1-\alpha_i}[f, g] := - f \cdot {}_{t_i}I_{b_i}^{1-\alpha_i}[g] + g \cdot {}_{a_i}I_{t_i}^{1-\alpha_i}[f], \quad i = 1, \dots, n,$$

function ξ is taken in $(t, y(t))$ and functions $\partial_j F$ are evaluated at

$$(y(t), \nabla_{I^{1-\alpha}}[y](t), \nabla[y](t), \nabla_{D^\alpha}[y](t), t)$$

for $j = 1, \dots, 3n$.

Corollary 4.66 *Assume $y \in C^1(\bar{\Omega}_n; \mathbb{R})$ with $_{a_i}I_{t_i}^{1-\alpha_i(\cdot)}[y], {}_{a_i}^C D_{t_i}^{\alpha_i(\cdot)}[y] \in C(\bar{\Omega}_n; \mathbb{R})$, $\alpha_i : [0, b_i - a_i] \to [0, 1]$, and that*

$$F(y(t), \nabla_I[y](t), \nabla[y](t), \nabla_D[y](t), t) = F(\hat{y}(t), \nabla_I[\hat{y}](t), \nabla[\hat{y}](t), \nabla_D[\hat{y}](t), t),$$

where $\hat{y}$ is the family (4.71) such that $t \mapsto \xi(t, y(t)) \in C^1(\bar{\Omega}_n; \mathbb{R})$ with $_{a_i}I_{t_i}^{1-\alpha_i(\cdot)}$ $[\tau \mapsto \xi(\tau, y(\tau))], {}_{a_i}^C D_{t_i}^{\alpha_i(\cdot)}[\tau \mapsto \xi(\tau, y(\tau))] \in C(\bar{\Omega}_n; \mathbb{R})$. Then,

$$\sum_{i=1}^n \left(\mathbf{I}_i^{1-\alpha_i(\cdot,\cdot)}\left[\xi, \partial_{1+i} F\right] + \frac{\partial}{\partial t_i}(\xi \cdot \partial_{1+n+i} F) + \mathbf{D}_i^{\alpha_i(\cdot,\cdot)}\left[\xi, \partial_{1+2n+i} F\right]\right) = 0$$

along any solution of the Euler–Lagrange equation (4.63), where

$$\mathbf{D}_i^{\alpha_i(\cdot,\cdot)}[f, g] := - f \cdot {}_{t_i}D_{b_i}^{\alpha_i(\cdot,\cdot)}[g] + g \cdot {}_{a_i}^C D_{t_i}^{\alpha_i(\cdot,\cdot)}[f],$$

$$\mathbf{I}_i^{1-\alpha_i(\cdot,\cdot)}[f, g] := - f \cdot {}_{t_i}I_{b_i}^{1-\alpha_i(\cdot,\cdot)}[g] + g \cdot {}_{a_i}I_{t_i}^{1-\alpha_i(\cdot,\cdot)}[f], \quad i = 1, \dots, n,$$

function ξ is taken in $(t, y(t))$ and functions $\partial_j F$ are evaluated at

$$(y(t), \nabla_I[y](t), \nabla[y](t), \nabla_D[y](t), t)$$

for $j = 1, \dots, 3n$.

4.8 Conclusion

In this chapter we unified, subsumed, and significantly extended the necessary optimality conditions available in the literature of the fractional calculus of variations. It should be mentioned, however, that since fractional operators are nonlocal, it can be extremely challenging to find analytical solutions to fractional problems of the calculus of variations and, in many cases, solutions may not exist. Here, we gave several examples with analytic solutions, and many more can be found borrowing different kernels from the book (Polyanin and Manzhirov 1998). On the other hand, one can easily choose examples for which the fractional Euler–Lagrange differential equations are hard to solve, and in that case one needs to use numerical methods (Almeida et al. 2012, 2015). However, in the absence of existence, the necessary conditions for extremality are vacuous: one cannot characterize an entity that does not exist in the first place. For solving a problem of the fractional calculus of variations one should proceed along the following three steps: (i) first, prove that a solution to the problem exists; (ii) second, verify the applicability of necessary optimality conditions; (iii) finally, apply the necessary conditions which identify the extremals (the candidates). Further elimination, if necessary, identifies the minimizer(s) of the problem. All three steps in the above procedure are crucial. As mentioned by Young in Young (1969), the calculus of variations has born from the study of necessary optimality conditions, but any such theory is "naive" until the existence of minimizers is verified. Therefore, in the next chapter, we shall follow the direct approach, first proving that a solution exists and next finding candidates with the help of the necessary optimality conditions.

References

Agrawal OP (2006) Fractional variational calculus and the transversality conditions. J Phys A Math Gen 39(33):10375–10384

Agrawal OP (2007) Generalized Euler-Lagrange equations and transversality conditions for FVPs in terms of the Caputo derivative. J Vib Control 13(9–10):1217–1237

Agrawal OP (2010) Generalized variational problems and Euler-Lagrange equations. Comput Math Appl 59(5):1852–1864

Almeida R, Malinowska AB, Torres DFM (2010) A fractional calculus of variations for multiple integrals with application to vibrating string. J Math Phys 51(3):033503, 12 pp

Almeida R, Pooseh S, Torres DFM (2012) Fractional variational problems depending on indefinite integrals. Nonlinear Anal 75(3):1009–1025

Almeida R, Pooseh S, Torres DFM (2015) Computational methods in the fractional calculus of variations. Imperial College Press, London

Almeida R, Torres DFM (2009a) Holderian variational problems subject to integral constraints. J Math Anal Appl 359(2):674–681

Almeida R, Torres DFM (2009b) Isoperimetric problems on time scales with nabla derivatives. J Vib Control 15(6):951–958

Baleanu D, Muslih IS (2005) Lagrangian formulation of classical fields within Riemann-Liouville fractional derivatives. Phys Scr 72(2–3):119–121

Blasjo V (2005) The isoperimetric problem. Amer Math Mon 112(6):526–566

Camargo RF, Chiacchio AO, Charnet R, Capelas de Oliveira E (2009) Solution of the fractional Langevin equation and the Mittag-Leffler functions. J Math Phys 6:063507, 8 pp

Cresson J (2007) Fractional embedding of differential operators and Lagrangian systems. J Math Phys 48(3):033504, 34 pp

Curtis JP (2004) Complementary extremum principles for isoperimetric optimization problems. Optim Eng 5(4):417–430

Dacorogna B, (2004) Introduction to the calculus of variations. Translated from the 1992 French original. Imperial College Press, London

Evans LC (2010) Partial differential equations, vol 19. 2nd edn. Graduate studies in mathematics. American Mathematical Society, Providence

Ferreira RAC, Torres DFM (2010) Isoperimetric problems of the calculus of variations on time scales. In: Leizarowitz A, Mordukhovich BS, Shafrir I, Zaslavski AJ (eds) Nonlinear analysis and optimization II. Contemporary mathematics. American Mathematical Society, Providence, pp 123–131

Frederico GSF, Torres DFM (2008) Fractional conservation laws in optimal control theory. Nonlinear Dyn 53(3):215–222

Frederico GSF, Torres DFM (2010) Fractional Noether's theorem in the Riesz-Caputo sense. Appl Math Comput 217(3):1023–1033

Gelfand IM, Fomin SV (2000) Calculus of variations. Dover Publications Inc, New York

Giaquinta M, Hildebrandt S (2004) Calculus of variations I. Springer, Berlin

Herrera L, Nunez L, Patino A, Rago H (1986) A variational principle and the classical and quantum mechanics of the damped harmonic oscillator. Am J Phys 54(3):273–277

Jost J, Li-Jost X (1998) Calculus of variations. Cambridge University Press, Cambridge

Kilbas AA, Srivastava HM, Trujillo JJ (2006) Theory and applications of fractional differential equations, vol 204. North-Holland mathematics studies. Elsevier, Amsterdam

Malinowska AB (2012) A formulation of the fractional Noether-type theorem for multidimensional Lagrangians. Appl Math Lett 25(11):1941–1946

Malinowska AB (2013) On fractional variational problems which admit local transformations. J Vib Control 19(8):1161–1169

Malinowska AB, Torres DFM (2012) Introduction to the fractional calculus of variations. Imperial College Press, London

Noether E (1918) Invariante Variationsprobleme. Nachr v d Ges d Wiss zu Göttingen, pp 235–257

Odzijewicz T (2013) Variable order fractional isoperimetric problem of several variables. Advances in the theory and applications of non-integer order systems 257:133–139

Odzijewicz T, Malinowska AB, Torres DFM (2012a) Generalized fractional calculus with applications to the calculus of variations. Comput Math Appl 64(10):3351–3366

Odzijewicz T, Malinowska AB, Torres DFM (2012b) Fractional variational calculus with classical and combined Caputo derivatives. Nonlinear Anal 75(3):1507–1515

Odzijewicz T, Malinowska AB, Torres DFM (2012c) Fractional calculus of variations in terms of a generalized fractional integral with applications to physics. Abstr Appl Anal 2012(871912):24

Odzijewicz T, Malinowska AB, Torres DFM (2012d) Green's theorem for generalized fractional derivatives. In: Chen W, Sun HG, Baleanu D (eds) Proceedings of FDA'2012, the 5th symposium on fractional differentiation and its applications, 14–17 May 2012, Hohai University, Nanjing, China. Paper #084

Odzijewicz T, Malinowska AB, Torres DFM (2012e) Variable order fractional variational calculus for double integrals. In: Proceedings of the IEEE conference on decision and control 6426489:6873–6878

Odzijewicz T, Malinowska AB, Torres DFM (2013a) Fractional variational calculus of variable order. Advances in harmonic analysis and operator theory, Operator theory: advances and applications, vol 229. Birkhäuser, Basel, pp 291–301

Odzijewicz T, Malinowska AB, Torres DFM (2013b) Green's theorem for generalized fractional derivative. Fract Calc Appl Anal 16(1):64–75

Odzijewicz T, Malinowska AB, Torres DFM (2013c) Fractional calculus of variations of several independent variables. Eur Phys J Spec Top 222(8):1813–1826

Odzijewicz T, Torres DFM (2011) Fractional calculus of variations for double integrals. Balkan J Geom Appl 16(2):102–113

Odzijewicz T, Torres DFM (2012) Calculus of variations with classical and fractional derivatives. Math Balkanica 26(1–2):191–202

Polyanin AD, Manzhirov AV (1998) Handbook of integral equations. CRC, Boca Raton

van Brunt B (2004) The calculus of cariations. Springer, New York

Young LC (1969) Lectures on the calculus of variations and optimal control theory. Foreword by Fleming WH, Saunders, Philadelphia

Chapter 5
Direct Methods in Fractional Calculus of Variations

Abstract In this chapter, under assumptions of regularity, convexity and coercivity, we obtain sufficient conditions ensuring the existence of minimizers for functionals with a Lagrangian depending on generalized fractional derivatives and integrals. Necessary optimality conditions of Euler–Lagrange type are also given.

Keywords Existence of minimizers · Regularity · Convexity · Coercivity · Tonelli-type theorem · Fractional Euler–Lagrange equation

In contrast with the standard approach presented in Chap. 4, we now use direct methods to address the problem of finding minima to generalized fractional functionals. First, we prove the existence of solutions in an appropriate space of functions and under suitable assumptions of regularity, coercivity, and convexity. Next, we proceed with the application of an optimality condition, and finish examining the candidates to arrive to solution. For the original paper where the method is developed, in the case of Riemann–Liouville fractional derivatives, we refer the reader to Bourdin (2013). The results presented in this chapter can be found in the paper (Bourdin et al. 2014) and are part of the Ph.D. thesis (Odzijewicz 2013). For an important particular case see also Bourdin et al. (2013).

Let us briefly describe the main contents of the chapter. In Sect. 5.1, we prove a Tonelli-type theorem ensuring existence of minimizers for generalized fractional functionals. We also give sufficient conditions for a regular Lagrangian and for a coercive functional. Section 5.2 is devoted to a necessary optimality condition for minimizers. In the last Sect. 5.3, we improve our results assuming more regularity of the Lagrangian and generalized fractional operators.

5.1 Existence of a Minimizer for a Generalized Functional

Let us recall that $1 < p, q < \infty$ and $\frac{1}{p} + \frac{1}{q} = 1$. In this section, our aim is to give sufficient conditions ensuring the existence of a minimizer for the following generalized Lagrangian functional:

© The Author(s) 2015

A.B. Malinowska et al., *Advanced Methods in the Fractional Calculus of Variations*, SpringerBriefs in Applied Sciences and Technology, DOI 10.1007/978-3-319-14756-7_5

$$\mathcal{I} : \mathcal{A} \longrightarrow \mathbb{R}$$

$$y \longmapsto \int_a^b F(y(t), K_P[y](t), \dot{y}(t), B_P[y](t), t) \, dt,$$

where $\mathcal{A}$ is a weakly closed subset of $W^{1,p}(a, b; \mathbb{R})$, $\dot{y}$ denotes the derivative of y, K_P is the generalized fractional integral with a kernel in $L^q(\Delta; \mathbb{R})$, $B_P = K_P \circ \frac{d}{dt}$ is the generalized fractional Caputo derivative, $P = \langle a, t, b, \lambda, \mu \rangle$ is a set of parameters, and F is a Lagrangian of class C^1:

$$F : \qquad \mathbb{R}^4 \times [a, b] \longrightarrow \mathbb{R}$$
$$(x_1, x_2, x_3, x_4, t) \longmapsto F(x_1, x_2, x_3, x_4, t).$$

5.1.1 A Tonelli-Type Theorem

In this section, we state a Tonelli-type theorem ensuring the existence of a minimizer for $\mathcal{I}$ with the help of general assumptions of regularity, coercivity, and convexity. These three hypothesis are usual in the classical case, see Dacorogna (2004).

Definition 5.1 A Lagrangian F is said to be regular if it satisfies:

- $t \mapsto F(y(t), K_P[y](t), \dot{y}(t), B_P[y](t), t) \in L^1(a, b; \mathbb{R})$;
- $t \mapsto \partial_1 F(y(t), K_P[y](t), \dot{y}(t), B_P[y](t), t) \in L^1(a, b; \mathbb{R})$;
- $t \mapsto \partial_2 F(y(t), K_P[y](t), \dot{y}(t), B_P[y](t), t) \in L^P(a, b; \mathbb{R})$;
- $t \mapsto \partial_3 F(y(t), K_P[y](t), \dot{y}(t), B_P[y](t), t) \in L^q(a, b; \mathbb{R})$;
- $t \mapsto \partial_4 F(y(t), K_P[y](t), \dot{y}(t), B_P[y](t), t) \in L^P(a, b; \mathbb{R})$,

for any $y \in W^{1,p}(a, b; \mathbb{R})$.

Definition 5.2 Functional $\mathcal{I}$ is said to be coercive on $\mathcal{A}$ if it satisfies:

$$\lim_{\substack{\|y\|_{W^{1,p}} \to \infty \\ y \in \mathcal{A}}} \mathcal{I}(y) = +\infty.$$

We are now in position to state the following result.

Theorem 5.3 (Tonelli-type Theorem) *Let us assume that*

- *F is regular;*
- *$\mathcal{I}$ is coercive on $\mathcal{A}$;*
- *$F(\cdot, t)$ is convex on $(\mathbb{R}^d)^4$ for any $t \in [a, b]$.*

Then, there exists a minimizer for $\mathcal{I}$.

Proof Since F is regular, for any $y \in \mathcal{A}$

$$t \mapsto F(y(t), K_P[y](t), \dot{y}(t), B_P[y](t), t) \in L^1(a, b; \mathbb{R})$$

and then $\mathcal{I}(y)$ exists in $\mathbb{R}$. Let us introduce a minimizing sequence $(y_n)_{n \in \mathbb{N}} \subset \mathcal{A}$ satisfying

$$\mathcal{I}(y_n) \longrightarrow \inf_{y \in \mathcal{A}} \mathcal{I}(y) < +\infty.$$

Since $\mathcal{I}$ is coercive, $(y_n)_{n \in \mathbb{N}}$ is bounded in $W^{1,p}(a, b; \mathbb{R})$. Since $W^{1,p}(a, b; \mathbb{R})$ is a reflexive Banach space, it exists a subsequence of $(y_n)_{n \in \mathbb{N}}$ weakly convergent in $W^{1,p}(a, b; \mathbb{R})$. In the following, we still denote this subsequence by $(y_n)_{n \in \mathbb{N}}$ and we denote by $\bar{y}$ its weak limit. Since $\mathcal{A}$ is a weakly closed subset of $W^{1,p}(a, b; \mathbb{R})$, $\bar{y} \in \mathcal{A}$. Finally, using the convexity of F, we have for any $n \in \mathbb{N}$:

$$\mathcal{I}(y_n) \geq \mathcal{I}(\bar{y}) + \int_a^b \partial_1 F \cdot (y_n(t) - \bar{y}(t)) + \partial_2 F \cdot (K_P[y_n](t) - K_P[\bar{y}](t))$$

$$+ \partial_3 F \cdot (\dot{y}_n(t) - \dot{\bar{y}}(t)) + \partial_4 F \cdot (B_P[y_n](t) - B_P[\bar{y}](t)) \, \mathrm{d}t, \qquad (5.1)$$

where $\partial_i F$ are taken in $(\bar{y}(t), K_P[\bar{y}](t), \dot{\bar{y}}(t), B_P[\bar{y}](t), t)$ for any $i = 1, 2, 3, 4$.

Now, from these four following facts:

- F is regular;
- $y_n \xrightarrow{W^{1,p}} \bar{y}$;
- K_P is linear bounded from $L^p(a, b; \mathbb{R})$ to $L^q(a, b; \mathbb{R})$;
- the compact embedding $W^{1,p}(a, b; \mathbb{R}) \hookrightarrow\!\!\!\!\to C([a, b]; \mathbb{R})$ holds;

one can easily conclude that

- $t \mapsto \partial_3 F(\bar{y}(t), K_P[\bar{y}](t), \dot{\bar{y}}(t), B_P[\bar{y}](t), t) \in L^q(a, b; \mathbb{R})$ and $\dot{y}_n \xrightarrow{L^p} \dot{\bar{y}}$;
- $t \mapsto \partial_4 F(\bar{y}(t), K_P[\bar{y}](t), \dot{\bar{y}}(t), B_P[\bar{y}](t), t) \in L^p(a, b; \mathbb{R})$ and $B_P[y_n] \xrightarrow{L^q} B_P[\bar{y}]$;
- $t \mapsto \partial_1 F(\bar{y}(t), K_P[\bar{y}](t), \dot{\bar{y}}(t), B_P[\bar{y}](t), t) \in L^1(a, b; \mathbb{R})$ and $y_n \xrightarrow{L^\infty} \bar{y}$;
- $t \mapsto \partial_2 F(\bar{y}(t), K_P[\bar{y}](t), \dot{\bar{y}}(t), B_P[\bar{y}](t), t) \in L^p(a, b; \mathbb{R})$ and $K_P[y_n] \xrightarrow{L^q} K[\bar{y}]$.

Finally, when n tends to ∞ in the inequality (5.1), we obtain

$$\inf_{y \in \mathcal{A}} \mathcal{I}(y) \geq \mathcal{I}(\bar{y}) \in \mathbb{R},$$

which completes the proof.

The first two hypothesis of Theorem 5.3 are very general. Consequently, in Sects. 5.1.2 and 5.1.3, we give concrete assumptions on F ensuring its regularity and the coercivity of $\mathcal{I}$.

The last hypothesis of convexity is strong. Nevertheless, from more regularity assumptions on F, on K_P and on B_P, we prove in Sect. 5.3 that we can provide versions of Theorem 5.3 with weaker convexity assumptions.

5.1.2 Sufficient Condition for Regular Lagrangians

In this section, we give a sufficient condition on F implying its regularity. First, for any $M \geq 1$, let us define the set $\mathscr{P}_M$ of maps P such that, for any $(x_1, x_2, x_3, x_4, t) \in \mathbb{R}^4 \times [a, b]$,

$$P(x_1, x_2, x_3, x_4, t) = \sum_{k=0}^{N} c_k(x_1, t) \, |x_2|^{d_{2,k}} \, |x_3|^{d_{3,k}} \, |x_4|^{d_{4,k}}$$

with $N \in \mathbb{N}$ and where, for any $k = 0, \dots, N$, $c_k : \mathbb{R} \times [a, b] \longrightarrow \mathbb{R}^+$ is continuous and $(d_{2,k}, d_{3,k}, d_{4,k}) \in [0, q] \times [0, p] \times [0, q]$ satisfies $d_{2,k} + (q/p)d_{3,k} + d_{4,k} \leq (q/M)$.

Remark 5.4 We call attention of the reader that subscript P in the notation of operators K_P, A_P and B_P means $P = \langle a, t, b, \lambda, \mu \rangle$.

The following lemma shows the interest of sets $\mathscr{P}_M$.

Lemma 5.5 *Let $M \geq 1$ and $P \in \mathscr{P}_M$. Then*

$$\forall y \in W^{1,p}(a, b; \mathbb{R}), \; t \mapsto P(y(t), K_P[y](t), \dot{y}(t), B_P[y](t), t) \in L^M(a, b; \mathbb{R}).$$

Proof For any $k = 0, \dots, N$, $c_k(u, t)$ is continuous and then it is in $L^\infty(a, b; \mathbb{R})$. We also have $|K_P[y]|^{d_{2,k}} \in L^{q/d_{2,k}}(a, b; \mathbb{R})$, $|\dot{y}|^{d_{3,k}} \in L^{p/d_{3,k}}(a, b; \mathbb{R})$ and $|B_P[y]|^{d_{4,k}} \in L^{q/d_{4,k}}(a, b; \mathbb{R})$. Consequently,

$$c_k(u, t) \, |K_P[y]|^{d_{2,k}} \, |\dot{y}|^{d_{3,k}} \, |B_P[y]|^{d_{4,k}} \in L^r(a, b; \mathbb{R}) \qquad (5.2)$$

with $r = q/(d_{2,k} + (q/p)d_{3,k} + d_{4,k}) \geq M$. The proof is complete.

Then, from this previous Lemma, one can easily obtain the following proposition.

Proposition 5.6 *Let us assume that there exist $P_0 \in \mathscr{P}_1$, $P_1 \in \mathscr{P}_1$, $P_2 \in \mathscr{P}_p$, $P_3 \in \mathscr{P}_q$ and $P_4 \in \mathscr{P}_p$ such that for any $(x_1, x_2, x_3, x_4, t) \in (\mathbb{R})^4 \times [a, b]$:*

- $|F(x_1, x_2, x_3, x_4, t)| \leq P_0(x_1, x_2, x_3, x_4, t);$
- $|\partial_1 F(x_1, x_2, x_3, x_4, t)| \leq P_1(x_1, x_2, x_3, x_4, t);$
- $|\partial_2 F(x_1, x_2, x_3, x_4, t)| \leq P_2(x_1, x_2, x_3, x_4, t);$
- $|\partial_3 F(x_1, x_2, x_3, x_4, t)| \leq P_3(x_1, x_2, x_3, x_4, t);$
- $|\partial_4 F(x_1, x_2, x_3, x_4, t)| \leq P_4(x_1, x_2, x_3, x_4, t).$

Then, F is regular.

This last proposition states that if the norms of F and of its partial derivatives are controlled from above by elements of $\mathscr{P}_M$, then F is regular. We will see some examples in Sect. 5.1.4.

5.1.3 Sufficient Condition for Coercive Functionals

The definition of coercivity for functional $\mathcal{I}$ is strongly dependent on the considered set $\mathcal{A}$. Consequently, in this section, we will consider an example of set $\mathcal{A}$ and we will give a sufficient condition on F ensuring the coercivity of $\mathcal{I}$ in this case. Precisely, let us consider $y_a \in \mathbb{R}$ and $\mathcal{A} = W_a^{1,p}(a, b; \mathbb{R})$, where $W_a^{1,p}(a, b; \mathbb{R}) := \{y \in W^{1,p}(a, b; \mathbb{R}), \ y(a) = y_a\}$. From the compact embedding $W^{1,p}(a, b; \mathbb{R}) \hookrightarrow\hookrightarrow C([a, b]; \mathbb{R})$, $W_a^{1,p}(a, b; \mathbb{R})$ is weakly closed in $W^{1,p}(a, b; \mathbb{R})$.

An important consequence of such a choice of set $\mathcal{A}$ is given by the following lemma.

Lemma 5.7 *There exist A_0, $A_1 \geq 0$ such that for any $y \in W_a^{1,p}(a, b; \mathbb{R})$:*

- $\|y\|_{L^\infty} \leq A_0 \|\dot{y}\|_{L^p} + A_1$;
- $\|K_P[y]\|_{L^q} \leq A_0 \|\dot{y}\|_{L^p} + A_1$;
- $\|B_P[y]\|_{L^q} \leq A_0 \|\dot{y}\|_{L^p} + A_1$.

Proof The last inequality comes from the boundedness of K_P. Let us consider the second one. For any $y \in W_a^{1,p}(a, b; \mathbb{R})$, we have $\|y\|_{L^p} \leq \|y - y_a\|_{L^p} + \|y_a\|_{L^p} \leq (b - a) \|\dot{y}\|_{L^p} + (b - a)^{1/p} |y_a|$. We conclude using again the boundedness of K_P. Now, let us consider the first inequality. For any $y \in W_a^{1,p}(a, b; \mathbb{R})$, we have $\|y\|_{L^\infty} \leq \|y - y_a\|_{L^\infty} + |y_a| \leq \|\dot{y}\|_{L^1} + |y_a| \leq (b-a)^{1/q} \|\dot{y}\|_{L^p} + |y_a|$. Finally, we have just to define A_0 and A_1 as the maxima of the appearing constants. The proof is complete.

Precisely, Lemma 5.7 states the *affine domination* of the term $\|\dot{y}\|_{L^p}$ on the terms $\|y\|_{L^\infty}$, $\|K_P[y]\|_{L^q}$ and $\|B_P[y]\|_{L^q}$ for any $y \in W_a^{1,p}(a, b; \mathbb{R})$. This characteristic of $W_a^{1,p}(a, b; \mathbb{R})$ leads us to give the following sufficient condition for a coercive functional $\mathcal{I}$.

Proposition 5.8 *Let us assume that for any $(x_1, x_2, x_3, x_4, t) \in \mathbb{R}^4 \times [a, b]$:*

$$F(x_1, x_2, x_3, x_4, t) \geq c_0 |x_3|^p + \sum_{k=1}^{N} c_k |x_1|^{d_{1,k}} |x_2|^{d_{2,k}} |x_3|^{d_{3,k}} |x_4|^{d_{4,k}},$$

with $c_0 > 0$ and $N \in \mathbb{N}$ and where, for any $k = 1, \ldots, N$, $c_k \in \mathbb{R}$ and $(d_{1,k}, d_{2,k}, d_{3,k}, d_{4,k}) \in \mathbb{R}^+ \times [0, q] \times [0, p] \times [0, q]$ satisfies:

$$d_{2,k} + (q/p)d_{3,k} + d_{4,k} \leq q \quad \text{and} \quad 0 \leq d_{1,k} + d_{2,k} + d_{3,k} + d_{4,k} < p.$$

Then, $\mathcal{I}$ is coercive on $W_a^{1,p}(a, b; \mathbb{R})$.

Proof Let us define $r_k = q/(d_{2,k} + d_{4,k} + (q/p)d_{3,k}) \geq 1$ and let r'_k, denote the adjoint of r_k i.e., $r'_k = \frac{r_k}{r_k-1}$. Using Hölder's inequality, one can easily prove that, for any $y \in W_a^{1,p}(a, b; \mathbb{R})$, we have

$$\mathcal{I}(y) \geq c_0 \|\dot{y}\|_{L^p} - (b-a)^{1/r'} \sum_{k=1}^{N} |c_k| \, \|y\|_{L^\infty}^{d_{1,k}} \|K_P[y]\|_{L^q}^{d_{2,k}} \|\dot{y}\|_{L^p}^{d_{3,k}} \|B_P[y]\|_{L^q}^{d_{4,k}} .$$

From the "affine domination" of the term $\|\dot{y}\|_{L^p}$ on the terms $\|y\|_{L^\infty}$, $\|K_P[y]\|_{L^q}$ and $\|B_P[y]\|_{L^q}$ for any $y \in W_a^{1,p}(a, b; \mathbb{R})$ (see Lemma 5.7) and from the assumption $0 \leq d_{1,k} + d_{2,k} + d_{3,k} + d_{4,k} < p$, we obtain that

$$\lim_{\substack{\|\dot{y}\|_{L^p} \to \infty \\ y \in W_a^{1,p}(a,b;\mathbb{R})}} \mathcal{I}(y) = +\infty.$$

Finally, from Lemma 5.7, we also have in $W_a^{1,p}(a, b; \mathbb{R})$:

$$\|\dot{y}\|_{L^p} \to \infty \iff \|y\|_{W^{1,p}} \to \infty.$$

Consequently, $\mathcal{I}$ is coercive on $W_a^{1,p}(a, b; \mathbb{R})$. The proof is complete. $\blacksquare$

In this section, we have studied the case where $\mathcal{A}$ is the weakly closed subset of $W^{1,p}(a, b; \mathbb{R})$ satisfying the initial condition $y(a) = y_a$. For other examples of set $\mathcal{A}$, let us note that all the results of this section are still valid when:

- $\mathcal{A}$ is a weakly closed subset of $W^{1,p}(a, b; \mathbb{R})$ satisfying a final condition in $t = b$;
- $\mathcal{A}$ is a weakly closed subset of $W^{1,p}(a, b; \mathbb{R})$ satisfying two boundary conditions in $t = a$ and in $t = b$.

5.1.4 Examples of Lagrangians

In this section, we give several examples of a convex Lagrangian F satisfying assumptions of Propositions 5.6 and 5.8. In consequence, they are examples of application of Theorem 5.3 in the case $\mathcal{A} = W_a^{1,p}(a, b; \mathbb{R})$.

Example 5.9 The most classical examples of a Lagrangian are the quadratic ones. Let us consider the following:

$$F(x_1, x_2, x_3, x_4, t) = c(t) + \frac{1}{2} \sum_{i=1}^{4} |x_i|^2 ,$$

where $c : [a, b] \to \mathbb{R}$ is of class C^1. One can easily check that F satisfies the assumptions of Propositions 5.6 and 5.8 with $p = q = 2$. Moreover, F satisfies the convexity hypothesis of Theorem 5.3. Consequently, one can conclude that there exists a minimizer of $\mathcal{I}$ defined on $W_a^{1,2}(a, b; \mathbb{R})$.

Example 5.10 Let us consider $p = q = 2$ and let us still denote by F the Lagrangian defined in Example 5.9. To obtain a more general example, one can define a Lagrangian F_1 from F as a time-dependent homothetic transformation and/or translation of its variables. Precisely,

$$F_1(x_1, x_2, x_3, x_4, t)$$
$$= F\left(c_1(t)x_1 + c_1^0(t), c_2(t)x_2 + c_2^0(t), c_3(t)x_3 + c_3^0(t), c_4(t)x_4 + c_4^0(t), t\right),$$
(5.3)

where $c_i : [a, b] \to \mathbb{R}$ and $c_i^0 : [a, b] \to \mathbb{R}$ are of class C^1 for any $i = 1, 2, 3, 4$. In this case, F_1 also satisfies convexity hypothesis of Theorem 5.3 and the assumptions of Proposition 5.6. Moreover, if c_3 is with values in $\mathbb{R}^+$, then F_1 also satisfies the assumptions of Proposition 5.8.

One should be careful: this last remark is not available in more general context. Precisely, if a general Lagrangian F satisfies the convexity hypothesis of Theorem 5.3 and assumptions of Propositions 5.6 and 5.8, then Lagrangian F_1 obtained by (5.3) also satisfies the convexity hypothesis of Theorem 5.3 and the assumptions of Proposition 5.6. Nevertheless, the assumption of Proposition 5.8 can be lost by this process.

Example 5.11 We can also study quasilinear examples given by Lagrangians of the type

$$F(x_1, x_2, x_3, x_4, t) = c(t) + \frac{1}{p}|x_3|^p + \sum_{i=1}^{4} f_i(t) \cdot x_i,$$

where $c : [a, b] \to \mathbb{R}$ and for any $i = 1, 2, 3, 4$, $f_i : [a, b] \to \mathbb{R}$ are of class C^1. In this case, F satisfies the assumptions of Propositions 5.6 and 5.8. Consequently, since F satisfies the convexity hypothesis of Theorem 5.3, one can conclude that there exists a minimizer of $\mathcal{I}$ defined on $W_a^{1,p}(a, b; \mathbb{R})$.

The most important constraint in order to apply Theorem 5.3 is the convexity hypothesis. This is the reason why the previous examples concern convex quasi–polynomial Lagrangians. Nevertheless, in Sect. 5.3, we are going to provide some improved versions of Theorem 5.3 with weaker convexity assumptions. This will be allowed by more regularity hypotheses on F and/or on K_P and B_P. We refer to Sect. 5.3 for more details.

5.2 Necessary Optimality Condition for a Minimizer

In this section, we assume additionally that

- F satisfies the assumptions of Proposition 5.6 (in particular, F is regular);
- $\mathcal{A}$ satisfies the following condition:

$$\forall y \in \mathcal{A}, \ \forall \eta \in \mathcal{C}_c^\infty, \ \exists 0 < \varepsilon \leq 1, \ \forall |h| \leq \varepsilon, \ y + h\eta \in \mathcal{A}. \qquad (5.4)$$

The assumption on $\mathcal{A}$ is satisfied if $\mathcal{A} + \mathcal{C}_c^\infty \subset \mathcal{A}$ (for example $\mathcal{A} = W_a^{1,p}(a, b; \mathbb{R})$ in Sect. 5.1.3). By $\mathcal{C}_c^\infty$ we denote the space of infinitely differentiable functions compactly supported in (a, b).

In the next theorem we will make use of the following Lemma.

Lemma 5.12 *Let* $M \geq 1$ *and* $P \in \mathcal{P}_M$. *Then, for any* $y \in \mathcal{A}$ *and any* $\eta \in \mathcal{C}_c^\infty$, *it exists* $g \in L^M(a, b; \mathbb{R}^+)$ *such that for any* $h \in [-\varepsilon, \varepsilon]$:

$$P(y + h\eta, K_P[y] + hK_P[\eta], \dot{y} + h\dot{\eta}, B_P[y] + hB_P[\eta], t) \leq g. \qquad (5.5)$$

Proof Indeed, for any $k = 0, \ldots, N$, for almost all $t \in (a, b)$ and for any $h \in [-\varepsilon, \varepsilon]$, we have

$$
\begin{aligned}
c_k(y(t) &+ h\eta(t), t) \, |K_P[y](t) + hK_P[\eta](t)|^{d_{2,k}} \, |\dot{y}(t) + h\dot{\eta}(t)|^{d_{3,k}} \\
&\times |B_P[y](t) + hB_P[\eta](t)|^{d_{4,k}} \\
\leq \bar{c}_k (&\underbrace{|K_P[y](t)|^{d_{2,k}} + |K_P[\eta](t)|^{d_{2,k}}}_{\in L^{q/d_{2,k}}(a,b;\mathbb{R})}) (\underbrace{|\dot{y}(t)|^{d_{3,k}} + |\dot{\eta}(t)|^{d_{3,k}}}_{\in L^{p/d_{3,k}}(a,b;\mathbb{R})}) \\
&\times \underbrace{(|B_P[y](t)|^{d_{4,k}} + |B_P[\eta](t)|^{d_{4,k}})}_{\in L^{q/d_{4,k}}(a,b;\mathbb{R})},
\end{aligned}
\qquad (5.6)
$$

where $\bar{c}_k = 2^{d_{2,k}+d_{3,k}+d_{4,k}} \max_{[a,b] \times [-\varepsilon, \varepsilon]} c_k(y(t) + h\eta(t), t)$ exists in $\mathbb{R}$ because c_k, y and η are continuous. Since $d_{2,k} + (q/p)d_{3,k} + d_{4,k} \leq (q/M)$, the right term in inequality (5.6) is in $L^M(a, b; \mathbb{R}^+)$ and is independent of h. The proof is complete.

Finally, from this previous Lemma, we can prove the following result.

Theorem 5.13 *Let us assume that* F *satisfies assumptions of Proposition 5.6,* $\mathcal{I}$ *is coercive and* $F(\cdot, t)$ *is convex on* $\mathbb{R}^4$ *for any* $t \in [a, b]$. *Then, the minimizer* $\bar{y} \in \mathcal{A}$ *of* $\mathcal{I}$ *(given by Theorem 5.3) satisfies the generalized Euler–Lagrange equation*

$$\frac{d}{dt}\left(\partial_3 F(\star_y)(t) + K_{P^*}[\partial_4 F(\star_y)(\tau)](t)\right) = \partial_1 F(\star_y)(t) + K_{P^*}[\partial_2 F(\star_y)(\tau)](t),$$
$$(\text{GEL})$$

for almost all $t \in (a, b)$, *where* $(\star_y)(t) = (y(t), K_P[y](t), \dot{y}(t), B_P[y](t), t)$.

Proof Since F satisfies the assumptions of Proposition 5.6, F is regular. Consequently, from Theorem 5.3, we know that $\mathcal{I}$ admits a minimizer $\bar{y} \in \mathcal{A}$ and then

$$\mathcal{I}(\bar{y}) \leq \mathcal{I}(\bar{y} + h\eta), \tag{5.7}$$

for any $|h| \leq \varepsilon$ and every $\eta \in \mathscr{C}_c^\infty$. Let us define the following map:

$$\phi_{\bar{y},\eta} : [-\varepsilon, \varepsilon] \longrightarrow \mathbb{R}$$

$$h \longmapsto \mathcal{I}(\bar{y} + h\eta) = \int_a^b \psi_{\bar{y},\eta}(t, h)\, dt,$$

where

$$\psi_{\bar{y},\eta}(t, h) := F(\bar{y}(t) + h\eta(t), K_P[\bar{y}](t) + hK_P[\eta](t), \dot{\bar{y}}(t)$$
$$+ h\dot{\eta}(t), B_P[\bar{y}](t) + hB_P[\eta](t), t).$$

First we want to prove that the following term:

$$\lim_{h \to 0} \frac{\mathcal{I}(\bar{y} + h\eta) - \mathcal{I}(\bar{y})}{h} = \lim_{h \to 0} \frac{\phi_{\bar{y},\eta}(h) - \phi_{\bar{y},\eta}(0)}{h} = \phi'_{\bar{y},\eta}(0) \tag{5.8}$$

exists in $\mathbb{R}$. In order to differentiate $\phi_{\bar{y},\eta}$, we use the theorem of differentiation under the integral sign. Indeed, we have for almost all $t \in (a, b)$ that $\psi_{\bar{y},\eta}(t, \cdot)$ is differentiable on $[-\varepsilon, \varepsilon]$ with

$$\frac{\partial \psi_{\bar{y},\eta}}{\partial h}(t, h) = \partial_1 F(\star_{\bar{y}+h\eta})(t) \cdot \eta(t) + \partial_2 F(\star_{\bar{y}+h\eta})(t) \cdot K_P[\eta](t)$$

$$+ \partial_3 F(\star_{\bar{y}+h\eta})(t) \cdot \dot{\eta}(t) + \partial_4 F(\star_{\bar{y}+h\eta})(t) \cdot B_P[\eta](t) \tag{5.9}$$

for all $h \in [-\varepsilon, \varepsilon]$. Since F satisfies the assumptions of Proposition 5.6, from Lemma 5.12 there exist $g_1 \in L^1(a, b; \mathbb{R}^+)$, $g_2 \in L^p(a, b; \mathbb{R}^+)$, $g_3 \in L^q(a, b; \mathbb{R}^+)$, and $g_4 \in L^p(a, b; \mathbb{R}^+)$ such that for any $h \in [-\varepsilon, \varepsilon]$ and for almost all $t \in (a, b)$:

$$\left| \frac{\partial \psi_{\bar{y},\eta}}{\partial h}(t, h) \right| \leq g_1(t) |\eta(t)| + g_2(t) |K_P[\eta](t)| + g_3(t) |\dot{\eta}(t)| + g_4(t) |B_P[\eta](t)|. \tag{5.10}$$

Since $\eta \in L^\infty(a, b; \mathbb{R})$, $K_P[\eta] \in L^q(a, b; \mathbb{R})$, $\dot{\eta} \in L^p(a, b; \mathbb{R})$, and $B_P[\dot{\eta}] \in L^q(a, b; \mathbb{R})$, we can conclude that the right term in inequality (5.10) is in $L^1(a, b; \mathbb{R}^+)$ and is independent of h. Consequently, we can use the theorem of differentiation under the integral sign and we obtain that $\phi_{\bar{y},\eta}$ is differentiable with

$$\forall h \in [-\varepsilon, \varepsilon], \ \phi'_{\bar{y},\eta}(h) = \int_a^b \frac{\partial \psi_{\bar{y},\eta}}{\partial h}(t, h)\, dt. \tag{5.11}$$

Furthermore, inequality (5.7) implies

$$\frac{d}{dh}\phi_{\bar{y},\eta}(h)\bigg|_{h=0} = 0,$$

what can be written as

$$\int_a^b \partial_1 F(\star_{\bar{y}})(t) \cdot \eta(t) + \partial_2 F(\star_{\bar{y}})(t) \cdot K_P[\eta](t) + \partial_3 F(\star_{\bar{y}})(t) \cdot \dot{\eta}(t)$$

$$+ \partial_4 F(\star_{\bar{y}})(t) \cdot B_P[\eta](t)\, dt = 0.$$

Moreover, applying integration by parts formula (4.5) one has

$$\int_a^b \left(\partial_1 F(\star_{\bar{y}})(t) + K_{P*}[\partial_2 F(\star_{\bar{y}})(\tau)]\right) \cdot \eta(t) + \left(\partial_3 F(\star_{\bar{y}})(t)\right.$$

$$\left. + K_{P*}[\partial_4 F(\star_{\bar{y}})(\tau)]\right) \cdot \dot{\eta}(t)\, dt = 0.$$

Then, taking an absolutely continuous anti-derivative w of $t \mapsto \partial_1 F(\star_{\bar{y}})(t) + K_{P*}[\partial_2 F(\star_{\bar{y}})(\tau)](t) \in L^1(a, b; \mathbb{R})$, we obtain using integration by parts that

$$\int_a^b \left(\partial_3 F(\star_{\bar{y}})(t) + K_{P*}[\partial_4 F(\star_{\bar{y}})(\tau)](t) - w(t)\right) \cdot \dot{\eta}(t)\, dt = 0.$$

From the du Bois Reymond Lemma there exists a constant $C \in \mathbb{R}$ such that for almost all $t \in (a, b)$, we have

$$\partial_3 F(\star_{\bar{y}})(t) + K^*[\partial_4 F(\star_{\bar{y}})(\tau)](t) = C + w(t). \tag{5.12}$$

Since the right term is absolutely continuous, we can differentiate it almost everywhere on (a, b). Finally, we obtain that $\bar{y}$ is a minimizer of $\mathcal{I}$. Then the following equation holds almost everywhere on (a, b):

$$\frac{d}{dt}\left(\partial_3 F(\star_{\bar{y}})(t) + K_{P*}[\partial_4 F(\star_{\bar{y}})(\tau)](t)\right) = \partial_1 F(\star_{\bar{y}})(t) + K_{P*}[\partial_2 F(\star_{\bar{y}})(\tau)](t). \tag{5.13}$$

The proof is complete.

We would like to remark that in the particular case, when $k^\alpha(t, \tau) = \frac{1}{\Gamma(1-\alpha)}(t - \tau)^{-\alpha}$, $k^\alpha \in L^q(\Delta; \mathbb{R})$ and $P = \langle a, t, b, 1, 0 \rangle$, then Sects. 5.1.1, 5.1.2, 5.1.3, 5.1.4, and 5.2 recover the case of the fractional variational functional

$$\mathcal{I} : \mathcal{A} \longrightarrow \mathbb{R}$$

$$y \longmapsto \int_a^b F(y(t), {}_aI_t^{1-\alpha}[y](t), \dot{y}(t), {}_a^C D_t^\alpha[y](t), t)\, dt$$

studied in Bourdin et al. (2013). Moreover, if we choose the kernel $k^\alpha(t, \tau) = \frac{1}{\Gamma(1-\alpha(t,\tau))}(t-\tau)^{-\alpha(t,\tau)}$, $k^\alpha \in L^q(\Delta; \mathbb{R})$, and the parameter set $P = \langle a, t, b, 1, 0 \rangle$, then Sects. 5.1.1, 5.1.2, 5.1.3, 5.1.4, and 5.2 restore the case of the variable-order fractional variational functional

$$\mathcal{I} : \mathcal{A} \longrightarrow \mathbb{R}$$

$$y \longmapsto \int_a^b F\left(y(t), {}_aI_t^{1-\alpha(\cdot,\cdot)}[y](t), \dot{y}(t), {}_a^C D_t^{\alpha(\cdot,\cdot)}[y](t), t\right)\, dt.$$

5.3 Some Improvements

In this section, we assume more regularity of the Lagrangian F and of the operators K_P and B_P. It allows to weaken the convexity assumption in Theorem 5.3 and/or the assumptions of Propositions 5.6 and 5.8.

5.3.1 A First Weaker Convexity Assumption

Let us assume that F satisfies the following condition:

$$(F(\cdot, x_2, x_3, x_4, t))_{(x_2, x_3, x_4, t) \in \mathbb{R}^3 \times [a,b]} \quad \text{is uniformly equicontinuous on } \mathbb{R}. \quad (5.14)$$

This condition has to be understood as:

$$\forall \varepsilon > 0, \exists \delta > 0, \forall (u, v) \in \mathbb{R}^2, \ |u - v| \le \delta \Longrightarrow \forall (x_2, x_3, x_4, t) \in \mathbb{R}^3 \times [a, b],$$
$$|F(u, x_2, x_3, x_4, t) - F(v, x_2, x_3, x_4, t)| \le \varepsilon. \quad (5.15)$$

For example, this condition is satisfied for a Lagrangian F with bounded $\partial_1 F$. In this case, we can prove the following improved version of Theorem 5.3:

Theorem 5.14 *Let us assume that*

- *F satisfies the condition given by (5.14);*
- *F is regular;*
- *$\mathcal{I}$ is coercive on $\mathcal{A}$;*
- *$F(x_1, \cdot, t)$ is convex on $\mathbb{R}^3$ for any $x_1 \in \mathbb{R}$ and for any $t \in [a, b]$.*

Then, there exists a minimizer for $\mathcal{I}$.

Proof Indeed, with the same proof of Theorem 5.3, we can construct a weakly convergent sequence $(y_n)_{n \in \mathbb{N}} \subset \mathcal{A}$ satisfying

$$y_n \xrightarrow{W^{1,p}} \bar{y} \in \mathcal{A} \text{ and } \mathcal{I}(y_n) \longrightarrow \inf_{y \in \mathcal{A}} \mathcal{I}(y) < +\infty.$$

Since the compact embedding $W^{1,p}(a, b; \mathbb{R}) \hookrightarrow C([a, b]; \mathbb{R})$ holds, we have $y_n \xrightarrow{C} \bar{y}$. Let $\varepsilon > 0$ and let us consider $\delta > 0$ given by (5.15). There exists $N \in \mathbb{N}$ such that for any $n \geq N$, $\|y_n - \bar{y}\|_\infty \leq \delta$. So, for any $n \geq N$ and for $t \in (a, b)$:

$$\left| F(y_n(t), K_P[y_n](t), \dot{y}_n(t), B_P[y_n](t), t) \right.$$

$$\left. - F(\bar{y}(t), K_P[y_n](t), \dot{y}_n(t), B_P[y_n](t), t) \right| \leq \varepsilon.$$

Consequently, for any $n \geq N$, we have

$$\mathcal{I}(y_n) \geq \int_a^b F(\bar{y}(t), K_P[y_n](t), \dot{y}_n(t), B_P[y_n](t), t) \, dt - (b - a)\varepsilon.$$

From the convexity hypothesis and using the same strategy as in the proof of Theorem 5.3, we have by passing to the limit on n

$$\inf_{y \in \mathcal{A}} \mathcal{I}(y) \geq \mathcal{I}(\bar{y}) - (b - a)\varepsilon.$$

The proof is complete since the previous inequality is true for any $\varepsilon > 0$.   ∎

Such an improvement allows to give examples of a Lagrangian F without convexity on its first variable. Taking inspiration from Example 5.9, we can provide the following example.

Example 5.15 Let us consider $p = 2$ and $\mathcal{A} = W_a^{1,2}(a, b; \mathbb{R})$. Let us consider

$$F(x_1, x_2, x_3, x_4, t) = f(x_1, t) + \frac{1}{2} \sum_{i=2}^{4} |x_i|^2,$$

for any $f : \mathbb{R} \times [a, b] \to \mathbb{R}$ of class C^1 with $\partial_1 f$ bounded (like sine and cosine functions). In this case, F satisfies the hypothesis of Theorem 5.14 and we can conclude with the existence of a minimizer of $\mathcal{I}$ defined on $\mathcal{A}$.

5.3.2 A Second Weaker Convexity Assumption

In this section, we assume that K_P is moreover a linear bounded operator from $C([a, b]; \mathbb{R})$ to $C([a, b]; \mathbb{R})$. For example, this condition is satisfied by Riemann–Liouville fractional integrals (see (Kilbas et al. 2006; Samko et al. 1993) for detailed proofs). We also assume that F satisfies the following condition:

$$(F(\cdot, \cdot, x_3, x_4, t))_{(x_3, x_4, t) \in \mathbb{R}^2 \times [a, b]} \text{ is uniformly equicontinuous on } \mathbb{R}^2. \quad (5.16)$$

This condition has to be understood as:

$$\forall \varepsilon > 0, \exists \delta > 0, \forall (u, v) \in \mathbb{R}^2, \forall (u_0, v_0) \in \mathbb{R}^2 \ |u - v| \le \delta, |u_0 - v_0| \le \delta$$
$$\implies \forall (x_3, x_4, t) \in \mathbb{R}^2 \times [a, b], |F(u, u_0, x_3, x_4, t) - F(v, v_0, x_3, x_4, t)| \le \varepsilon.$$
$$(5.17)$$

For example, this condition is satisfied for a Lagrangian F with bounded $\partial_1 F$ and bounded $\partial_2 F$. In this case, we can prove the following improved version of Theorem 5.3.

Theorem 5.16 *Let us assume that*

- *F satisfies the condition given by (5.16);*
- *F is regular;*
- *$\mathcal{I}$ is coercive on $\mathcal{A}$;*
- *$F(x_1, x_2, \cdot, t)$ is convex on $\mathbb{R}^2$ for any $(x_1, x_2) \in \mathbb{R}$ and for any $t \in [a, b]$.*

Then, there exists a minimizer for $\mathcal{I}$.

Proof Indeed, with the same proof of Theorem 5.3, we can construct a weakly convergent sequence $(y_n)_{n \in \mathbb{N}} \subset \mathcal{A}$ satisfying

$$y_n \xrightarrow{W^{1,p}} \bar{y} \in \mathcal{A} \text{ and } \mathcal{I}(y_n) \longrightarrow \inf_{y \in \mathcal{A}} \mathcal{I}(y) < +\infty.$$

Since the compact embedding $W^{1,p}(a, b; \mathbb{R}) \hookrightarrow C([a, b]; \mathbb{R})$ holds, we have $y_n \xrightarrow{C} \bar{y}$ and since K_P is continuous from $C([a, b]; \mathbb{R})$ to $C([a, b]; \mathbb{R})$, we have $K_P[\bar{y}_n] \xrightarrow{C} K_P[\bar{y}]$. Let $\varepsilon > 0$ and let us consider $\delta > 0$ given by (5.17). There exists $N \in \mathbb{N}$ such that for any $n \ge N$, $\|y_n - \bar{y}\|_\infty \le \delta$ and $\|K_P[y_n] - K_P[\bar{y}]\|_\infty \le \delta$. So, for any $n \ge N$ and for $t \in (a, b)$:

$$\left| F(y_n(t), K_P[y_n](t), \dot{y}_n(t), B_P[y_n](t), t) \right.$$
$$\left. - F(\bar{y}(t), K_P[\bar{y}](t), \dot{y}_n(t), B_P[y_n](t), t) \right| \le \varepsilon.$$

Consequently, for any $n \geq N$, we have

$$\mathcal{I}(y_n) \geq \int_a^b F(\bar{y}(t), K_P[\bar{y}](t), \dot{y}_n(t), B_P[y_n](t), t) \, dt - (b - a)\varepsilon.$$

From the convexity hypothesis and using the same strategy as in the proof of Theorem 5.3, we have by passing to the limit on n:

$$\inf_{y \in \mathcal{A}} \mathcal{I}(y) \geq \mathcal{I}(\bar{y}) - (b - a)\varepsilon.$$

The proof is complete since the previous inequality is true for any $\varepsilon > 0$.

Such an improvement allows to give examples of a Lagrangian F without convexity on its two first variables. Taking inspiration from Example 5.11, we can provide the following example.

Example 5.17 Let us consider

$$F(x_1, x_2, x_3, x_4, t) = c(t) \cos(x_1) \cdot \sin(x_2) + \frac{1}{p} |x_3|^p + f(t) \cdot x_4,$$

where $c : [a, b] \to \mathbb{R}$, $f : [a, b] \to \mathbb{R}$ are of class $C^1([a, b]; \mathbb{R})$. In this case, one can prove that F satisfies all hypothesis of Theorem 5.16 and then, we can conclude with the existence of a minimizer of $\mathcal{I}$ defined on $W_a^{1,p}(a, b; \mathbb{R})$ for any $1 < p < \infty$ and $1 < q < \infty$.

5.4 Conclusion

In Bourdin et al. (2013) existence results for fractional variational problems containing Caputo derivatives were given. This chapter extends those results to any linear operator K_P bounded from the space $L^p(a, b; \mathbb{R})$ to $L^q(a, b; \mathbb{R})$, having in mind that $B_P := K_P \circ \frac{d}{dt}$.

References

Bourdin L (2013) Existence of a weak solution for fractional Euler-Lagrange equations. J Math Anal Appl 399(1):239–251

Bourdin L, Odzijewicz T, Torres DFM (2013) Existence of minimizers for fractional variational problems containing Caputo derivatives. Adv Dyn Syst Appl 8(1):3–12

Bourdin L, Odzijewicz T, Torres DFM (2014) Existence of minimizers for generalized Lagrangian functionals and a necessary optimality condition—application to fractional variational problems. Differ Integral Equ 27(7–8):743–766

Dacorogna B, (2004) Introduction to the calculus of variations. Translated from the 1992 French original. Imperial College Press, London

Kilbas AA, Srivastava HM, Trujillo JJ (2006) Theory and applications of fractional differential equations, vol 204. North-Holland Mathematics Studies. Elsevier, Amsterdam

Odzijewicz T (2013) Generalized fractional calculus of variations. PhD Thesis, University of Aveiro

Samko SG, Kilbas AA, Marichev OI (1993) Fractional integrals and derivatives. Translated from the 1987 Russian original. Gordon and Breach, Yverdon

Chapter 6
Application to the Sturm–Liouville Problem

Abstract We study the Sturm–Liouville eigenvalue problem with Caputo fractional derivatives and show that fractional variational principles are useful for proving existence of eigenvalues and eigenfunctions.

Keywords Fractional Sturm–Liouville problem · Eigenvalues · Eigenfunctions · Fractional Euler–Lagrange equation · Fractional variational calculus

In 1836–1837, the French mathematicians Sturm (1803–1853) and Liouville (1809–1855) published a series of articles initiating a new subtopic of mathematical analysis—the Sturm–Liouville theory. It deals with the general linear, second-order ordinary differential equation of the form

$$\frac{d}{dt}\left(p(t)\frac{dy}{dt}\right) + q(t)y = \lambda w(t)y, \tag{6.1}$$

where $t \in [a, b]$, and in any particular problem functions $p(t)$, $q(t)$, and $w(t)$ are known. In addition, certain boundary conditions are attached to Eq. (6.1). For specific choices of the boundary conditions, nontrivial solutions of (6.1) exist only for particular values of the parameter $\lambda = \lambda^{(m)}$, $m = 1, 2, \ldots$. Constants $\lambda^{(m)}$ are called eigenvalues and corresponding solutions $y^{(m)}(t)$ are called eigenfunctions. For a deeper discussion of the classical Sturm–Liouville theory, we refer the reader to Gelfand and Fomin (2000), van Brunt (2004). The results of this chapter can be found in the paper (Klimek et al. 2014) and are part of the PhD thesis (Odzijewicz 2013).

Recently, many researchers focused their attention on certain generalizations of the Sturm–Liouville problem. Namely, they are interested in equations of type (6.1), however, with fractional differential operators (see, e.g., (Al-Mdallal 2009, 2010; Klimek and Agrawal 2012, 2013a, b; Liu et al. 2012; Qi and Chen 2011)). In this chapter, we develop the Sturm–Liouville theory by studying the Sturm–Liouville eigenvalue problem with Caputo fractional derivatives. We show that fractional variational principles are useful for the approximation of eigenvalues and eigenfunctions. Traditional Sturm–Liouville theory does not depend upon the calculus

© The Author(s) 2015
A.B. Malinowska et al., *Advanced Methods in the Fractional Calculus
of Variations*, SpringerBriefs in Applied Sciences and Technology,
DOI 10.1007/978-3-319-14756-7_6

of variations, but stems from the theory of ordinary linear differential equations. However, the Sturm–Liouville eigenvalue problem is readily formulated as a constrained variational principle, and this formulation can be used to approximate the solutions. We emphasize that it has a special importance for the fractional Sturm–Liouville equation since fractional operators are nonlocal and it can be extremely challenging to find analytical solutions. Besides, allowing convenient approximations, many general properties of the eigenvalues can be derived using the variational principle.

6.1 Useful Lemmas

In this section, we present three lemmas that are used to prove existence of solutions for the fractional Sturm–Liouville problem.

Lemma 6.1 *Let $\alpha \in (0, 1)$ and function $\gamma \in C([a, b]; \mathbb{R})$. If*

$$\int_a^b \gamma(t) \frac{d}{dt} \left({}^C_a D_t^\alpha [h](t) \right) \, dt = 0$$

for each $h \in C^1([a, b]; \mathbb{R})$ such that $\frac{d}{dt} {}^C_a D_t^\alpha [h] \in C([a, b]; \mathbb{R})$ and boundary conditions

$$h(a) = {}_a I_t^{1-\alpha} h(b) = 0$$

and

$$ {}^C_a D_t^\alpha [h](t)|_{t=a} = {}^C_t D_b^\alpha [h](t)|_{t=b} = 0$$

are fulfilled, then $\gamma(t) = c_0 + c_1 t$, where c_0, c_1 are some real constants.

Proof Let us define function h as follows:

$$h(t) := {}_a I_t^{1+\alpha} [\gamma(\tau) - c_0 - c_1\tau](t) \tag{6.2}$$

with constants fixed by the conditions

$$ {}_a I_t^2 [\gamma(\tau) - c_0 - c_1\tau](t)|_{t=b} = 0, \tag{6.3}$$

$$ {}_a I_t^1 [\gamma(\tau) - c_0 - c_1\tau](t)|_{t=b} = 0. \tag{6.4}$$

Observe that function h is continuous and fulfills the boundary conditions

$$h(a) = 0 \quad {}_a I_t^{1-\alpha} [h](t)|_{t=b} = {}_a I_t^2 [\gamma(\tau) - c_0 - c_1\tau](t)|_{t=b} = 0$$

and

$$ {}^C_a D_t^\alpha [h](t)|_{t=a} = {}_a D_t^\alpha [h](t)|_{t=a} = \frac{d}{dt} {}_a I_t^2 [\gamma(\tau) - c_0 - c_1\tau](t)|_{t=a}$$

$$= {}_a I_t^1 \left[\gamma(\tau) - c_0 - c_1 \tau \right](t)|_{t=a} = 0,$$

$$_a^C D_t^\alpha [h](t)|_{t=b} = {}_a D_t^\alpha [h](t)|_{t=b} = \frac{d}{dt} {}_a I_t^2 \left[\gamma(\tau) - c_0 - c_1 \tau \right](t)|_{t=b}$$

$$= {}_a I_t^1 \left[\gamma(\tau) - c_0 - c_1 \tau \right](t)|_{t=b} = 0.$$

In addition,

$$t \mapsto h'(t) = {}_a I_t^\alpha [\gamma(\tau) - c_0 - c_1 \tau](t) \in C([a, b]; \mathbb{R}),$$

$$t \mapsto \frac{d}{dt} {}_a^C D_t^\alpha [h](t) = \gamma(t) - c_0 - c_1 t \in C([a, b]; \mathbb{R}).$$

We also have

$$\int_a^b (\gamma(t) - c_0 - c_1 t) \frac{d}{dt} \left({}_a^C D_t^\alpha [h](t) \right) dt$$

$$= \int_a^b (-c_0 - c_1 t) \frac{d}{dt} \left({}_a^C D_t^\alpha [h](t) \right) dt$$

$$= -c_0 \cdot {}_a^C D_t^\alpha [h](t)|_{t=a}^{t=b} - c_1 t \cdot {}_a^C D_t^\alpha [h](t)|_{t=a}^{t=b} + c_1 \cdot {}_a I_t^{1-\alpha} [h](t)|_{t=a}^{t=b} = 0.$$

On the other hand,

$$\frac{d}{dt} \left({}_a^C D_t^\alpha [h](t) \right) = \frac{d}{dt} {}_a^C D_t^\alpha \left[{}_a I_t^{1+\alpha} [\gamma(\tau) - c_0 - c_1 \tau](s) \right](t) = \gamma(t) - c_0 - c_1 t$$

and

$$0 = \int_a^b (\gamma(t) - c_0 - c_1 t) \frac{d}{dt} \left({}_a^C D_t^\alpha [h](t) \right) dt = \int_a^b (\gamma(t) - c_0 - c_1 t)^2 \, dt.$$

Thus function γ is

$$\gamma(t) = c_0 + c_1 t.$$

The proof is complete.

Lemma 6.2 Let $\alpha \in \left(\frac{1}{2}, 1 \right)$, $\gamma \in C([a, b]; \mathbb{R})$ and ${}_a D_t^{1-\alpha} [\gamma] \in L^2(a, b; \mathbb{R})$. If

$$\int_a^b \gamma(t) \frac{d}{dt} \left({}_a^C D_t^\alpha [h](t) \right) dt = 0$$

for each $h \in C^1([a, b]; \mathbb{R})$ such that $h'' \in L^2(a, b; \mathbb{R})$, $\frac{d}{dt} {}_a^C D_t^\alpha[h] \in C([a, b]; \mathbb{R})$ and boundary conditions

$$h(a) = {}_a I_t^{1-\alpha}[h](b) = 0, \tag{6.5}$$

$$ {}_a^C D_t^\alpha[h](t)|_{t=a} = {}_a^C D_t^\alpha[h](t)|_{t=b} = 0 \tag{6.6}$$

are fulfilled, then $\gamma(t) = c_0 + c_1 t$, where c_0, c_1 are some real constants.

Proof We define function h as in the proof of Lemma 6.1:

$$h(t) := {}_a I_t^{1+\alpha} [\gamma(\tau) - c_0 - c_1 \tau](t) \tag{6.7}$$

with constants fixed by the conditions (6.3) and (6.4). The proof of the lemma is analogous to that of the Lemma 6.1. In addition, for the second-order derivative we have

$$
\begin{aligned}
h''(t) &= \frac{d}{dt} {}_a I_t^\alpha [\gamma(\tau) - c_0 - c_1 \tau](t) \\
&= {}_a D_t^{1-\alpha} [\gamma(\tau) - c_0 - c_1 \tau](t) \\
&= {}_a D_t^{1-\alpha}[\gamma](t) - (c_0 + c_1 a)\frac{(t-a)^{\alpha-1}}{\Gamma(\alpha)} - c_1 \frac{(t-a)^\alpha}{\Gamma(\alpha+1)}.
\end{aligned}
$$

Let us observe that for $\alpha > 1/2$:

$$t \mapsto \frac{(t-a)^{\alpha-1}}{\Gamma(\alpha)} \in L^2(a, b; \mathbb{R}),$$

$$t \mapsto \frac{(t-a)^\alpha}{\Gamma(\alpha+1)} \in C([a, b]; \mathbb{R}) \subset L^2(a, b; \mathbb{R}).$$

Thus, we conclude that $h'' \in L^2(a, b; \mathbb{R})$ and function h constructed in this proof fulfills all the assumptions of Lemma 6.2. The remaining part of the proof is analogous to that for Lemma 6.1.

Lemma 6.3 (a) *Let* $\alpha \in (\frac{1}{2}, 1)$, *functions* $\gamma_j \in C([a, b]; \mathbb{R})$, $j = 1, 2, 3$ *and* ${}_a D_t^{1-\alpha}[\gamma_3] \in L^2(a, b; \mathbb{R})$. *If*

$$\int_a^b \left(\gamma_1(t)h(t) + \gamma_2(t){}_a^C D_t^\alpha[h](t) + \gamma_3(t)\frac{d}{dt}\left({}_a^C D_t^\alpha[h](t)\right) \right) dt = 0 \tag{6.8}$$

for each $h \in C^1([a, b]; \mathbb{R})$, *such that* $h'' \in L^2(a, b; \mathbb{R})$ *and* $\frac{d}{dt} {}_a^C D_t^\alpha[h] \in C([a, b]; \mathbb{R})$, *fulfilling boundary conditions*

$$h(a) = {}_aI_t^{1-\alpha}[h](b) = 0, \tag{6.9}$$

$${}_a^CD_t^\alpha[h](t)|_{t=a} = {}_a^CD_t^\alpha[h](t)|_{t=b} = 0, \tag{6.10}$$

then $\gamma_3 \in C^1([a, b]; \mathbb{R})$.

(b) *Let* $\alpha \in \left(\frac{1}{2}, 1\right)$ *and functions* $\gamma_1, \gamma_2 \in C([a, b]; \mathbb{R})$. *If*

$$\int_a^b \left(\gamma_1(t)h(t) + \gamma_2(t){}_a^CD_t^\alpha[h](t)\right) dt = 0 \tag{6.11}$$

for each $h \in C^1([a, b]; \mathbb{R})$, *such that* $h'' \in L^2(a, b; \mathbb{R})$ *and* $\frac{d}{dt}{}_a^CD_t^\alpha[h] \in C([a, b]; \mathbb{R})$, *fulfilling boundary conditions* (6.9) *and* (6.10), *then*

$$-\gamma_1(t) - {}_t^CD_b^\alpha[\gamma_2](t) = 0.$$

Proof Observe that integral (6.8) can be rewritten as

$$\int_a^b \left(\gamma_1(t)h(t) + \gamma_2(t){}_a^CD_t^\alpha[h](t) + \gamma_3(t)\frac{d}{dt}{}_a^CD_t^\alpha[h](t)\right) dt$$

$$= \int_a^b \left(-\left({}_aI_t^1 \circ {}_tI_b^\alpha\right)[\gamma_1](t) - {}_aI_t^1[\gamma_2](t) + \gamma_3(t)\right)\frac{d}{dt}{}_a^CD_t^\alpha[h](t) \, dx = 0$$

due to the fact that relations

$$\left({}_aI_t^\alpha \circ {}_tI_b^1 \circ \frac{d}{dt}{}_a^CD_t^\alpha\right)[h](t) = h(t)$$

and

$$\left({}_tI_b^1 \circ \frac{d}{d\tau}{}_a^CD_t^\alpha\right)[h](t) = -{}_a^CD_t^\alpha[h](t)$$

are valid because function h fulfills boundary conditions (6.9) and (6.10). Denote

$$\gamma(t) := -\left({}_aI_t^1 \circ {}_tI_b^\alpha\right)[\gamma_1](t) - {}_aI_t^1[\gamma_2](t) + \gamma_3(t).$$

It is clear that $\gamma \in C([a, b]; \mathbb{R})$ and ${}_aD_t^{1-\alpha}[\gamma] \in L^2(a, b; \mathbb{R})$. Thus, according to Lemma 6.2, there exist constants c_0 and c_1 such that

$$-\left({}_aI_t^1 \circ {}_tI_b^\alpha\right)[\gamma_1](t) - {}_aI_t^1[\gamma_2](t) + \gamma_3(t) = c_0 + c_1t.$$

Let us note that function γ_3 is

$$\gamma_3(t) = \left({}_a I_t^1 \circ {}_t I_b^\alpha\right) [\gamma_1](t) + {}_a I_t^1 [\gamma_2](t) + c_0 + c_1 t.$$

Hence its first order derivative is continuous in $[a, b]$ and $\gamma_3 \in C^1([a, b]; \mathbb{R})$. The proof of part (b) is similar. We write integral (6.11) as follows:

$$\int_a^b \left(\gamma_1(t)h(t) + \gamma_2(t) {}_a^C D_t^\alpha[h](t)\right) dt$$

$$= \int_a^b \left(-\left({}_a I_t^1 \circ {}_t I_b^\alpha\right) [\gamma_1](t) - {}_a I_t^1 [\gamma_2](t)\right) \frac{d}{dt} {}_a^C D_t^\alpha[h](t) \, dt = 0.$$

The function in brackets is continuous in $[a, b]$,

$${}_a D_t^{1-\alpha}\left[-\left({}_a I_t^1 \circ {}_t I_b^\alpha\right) [\gamma_1](\tau) - {}_a I_t^1 [\gamma_2](\tau)\right](t)$$
$$= -\left({}_a I_t^\alpha \circ {}_t I_b^\alpha\right) [\gamma_1](t) - {}_a I_t^\alpha [\gamma_2](t),$$

and

$$-\left({}_a I_t^\alpha \circ {}_t I_b^\alpha\right) [\gamma_1] - {}_a I_t^\alpha [\gamma_2] \in C([a, b]; \mathbb{R}) \subset L^2(a, b; \mathbb{R}),$$

so we again can apply Lemma 6.2 and obtain that there exist constants c_0 and c_1 such that

$$\left({}_a I_t^1 \circ {}_t I_b^\alpha\right) [\gamma_1](t) + {}_a I_t^1 [\gamma_2](t) = c_0 + c_1 t.$$

Thus, functions $\gamma_{1,2}$ fulfill equation ${}_t^C D_b^\alpha[\gamma_2](t) + \gamma_1(t) = 0$.

6.2   The Fractional Sturm–Liouville Problem

The crucial idea in the proof of main result of this chapter (Theorem 6.5) is to apply direct variational methods to the fractional Sturm–Liouville equation. Starting from the fractional Sturm–Liouville equation, the approach is to find an associated functional and to use this to find approximations to the minimizers, which are necessarily solutions to the original equation. In the case of the fractional Sturm–Liouville equation, an associated variational problem is the fractional isoperimetric problem, which is defined in the following way:

$$\min \mathcal{I}(y) = \int_a^b F\left(y(t), {}_a^C D_t^\alpha[y](t), t\right) dt, \tag{6.12}$$

subject to the boundary conditions

$$y(a) = y_a, \quad y(b) = y_b \tag{6.13}$$

and the isoperimetric constraint

$$\mathcal{J}(y) = \int_a^b G\left(y(t), {}_a^C D_t^\alpha[y](t), t\right) \, dt = \xi, \tag{6.14}$$

where $\xi \in \mathbb{R}$ is given, and

$$F : [a, b] \times \mathbb{R}^2 \longrightarrow \mathbb{R}$$
$$(y, u, t) \longmapsto F(y, u, t),$$

$$G : [a, b] \times \mathbb{R}^2 \longrightarrow \mathbb{R}$$
$$(y, u, t) \longmapsto G(y, u, t)$$

are functions of class C^1, such that $\frac{\partial F}{\partial u}$, $\frac{\partial G}{\partial u}$ have continuous ${}_tD_b^\alpha$ derivatives.

Theorem 6.4 (cf. Theorem 3.3 (Almeida and Torres 2011)) *If $\bar{y} \in C[a, b]$ with ${}_a^C D_t^\alpha[\bar{y}] \in C([a, b]; \mathbb{R})$ is a minimizer for problem (6.12)–(6.14), then there exists a real constant λ such that, for $H = F + \lambda G$, the Euler–Lagrange equation*

$$\frac{\partial H}{\partial y}(y(t), {}_a^C D_t^\alpha[y](t), t) + {}_tD_b^\alpha\left[\frac{\partial H}{\partial u}(y(t), {}_a^C D_t^\alpha[y](t), t)\right] = 0 \tag{6.15}$$

holds for $\bar{y}$, provided $\bar{y}$ is not an Euler–Lagrange extremal for G, that is,

$$\frac{\partial G}{\partial y}(\bar{y}(t), {}_a^C D_t^\alpha[\bar{y}](t), t) + {}_tD_b^\alpha\left[\frac{\partial G}{\partial u}(\bar{y}(t), {}_a^C D_t^\alpha[\bar{y}](t), t)\right] \neq 0.$$

6.2.1 Existence of Discrete Spectrum

We show that, similarly to the classical case, for the fractional Sturm–Liouville problem there exists an infinite monotonic increasing sequence of eigenvalues. Moreover, apart from multiplicative factors to each eigenvalue, there corresponds precisely one eigenfunction and eigenfunctions form an orthogonal set of solutions.

We shall use the following assumptions.

(H1) Let $\frac{1}{2} < \alpha < 1$ and p, q, w_α be given functions such that: p is of C^1 class and $p(t) > 0$; q, w_α are continuous, $w_\alpha(t) > 0$ and $(\sqrt{w_\alpha})'$ is Hölderian of order $\beta \leq \alpha - \frac{1}{2}$. Consider the fractional differential equation

$$\,^{C}_{t}D^{\alpha}_{b}\left[p(\tau)\,^{C}_{a}D^{\alpha}_{\tau}[y](\tau)\right](t)+q(t)y(t)=\lambda w_{\alpha}(t)y(t), \qquad (6.16)$$

that will be called the fractional Sturm–Liouville equation, subject to the boundary conditions

$$y(a)=y(b)=0. \qquad (6.17)$$

Theorem 6.5 *Under assumptions (H1), the fractional Sturm–Liouville problem (FSLP) (6.16) and (6.17) has an infinite increasing sequence of eigenvalues* $\lambda^{(1)}$, $\lambda^{(2)},\ldots$, *and to each eigenvalue* $\lambda^{(n)}$ *there corresponds an eigenfunction* $y^{(n)}$ *which is unique up to a constant factor. Furthermore, eigenfunctions* $y^{(n)}$ *form an orthogonal set of solutions.*

Proof The proof is similar in spirit to Gelfand and Fomin (2000) and will be divided into 6 steps. As in Gelfand and Fomin (2000), we shall derive a method for approximating both eigenvalues and eigenfunctions.

Step 1. We shall consider the problem of minimizing the functional

$$\mathcal{I}(y)=\int_{a}^{b}\left[p(t)(\,^{C}_{a}D^{\alpha}_{t}[y](t))^{2}+q(t)y^{2}(t)\right]\mathrm{d}t \qquad (6.18)$$

subject to an isoperimetric constraint

$$\mathcal{J}(y)=\int_{a}^{b}w_{\alpha}(t)y^{2}(t)\,\mathrm{d}t=1 \qquad (6.19)$$

and boundary conditions (6.17). First, let us point out that functional (6.18) is bounded from below. Indeed, as $p(t)>0$ we have

$$\mathcal{I}(y)=\int_{a}^{b}\left[p(t)(\,^{C}_{a}D^{\alpha}_{t}[y](t))^{2}+q(t)y^{2}(t)\right]\mathrm{d}t$$

$$\geq\min_{t\in[a,b]}\frac{q(t)}{w_{\alpha}(t)}\cdot\int_{a}^{b}w_{\alpha}(t)y^{2}(t)\,\mathrm{d}t=\min_{t\in[a,b]}\frac{q(t)}{w_{\alpha}(t)}=:M_{0}>-\infty.$$

From now on, for simplicity, we assume that $a=0$ and $b=\pi$. According to the Ritz method, we approximate solution of (6.17)–(6.19) using the following trigonometric function with coefficient depending on w_{α}:

$$y_{m}(t)=\frac{1}{\sqrt{w_{\alpha}}}\sum_{k=1}^{m}\beta_{k}\sin(kt). \qquad (6.20)$$

Observe that $y_m(0) = y_m(\pi) = 0$. Substituting (6.20) into (6.18) and (6.19) we obtain the problem of minimizing the function

$$
I(\beta_1, \ldots, \beta_m) = I([\beta])
$$

$$
= \sum_{k,j=1}^{m} \beta_k \beta_j \times \int_0^\pi \left[p(t) \left({}_0^C D_t^\alpha \left[\frac{\sin(k\tau)}{\sqrt{w_\alpha}} \right](t) \cdot {}_0^C D_t^\alpha \left[\frac{\sin(j\tau)}{\sqrt{w_\alpha}} \right](t) \right) \right.
$$

$$
\left. + \frac{q(t)}{w_\alpha(t)} \sin(kt) \sin(jt) \right] dt \tag{6.21}
$$

subject to the condition

$$
J(\beta_1, \ldots, \beta_m) = J([\beta]) = \frac{\pi}{2} \sum_{k=1}^{m} (\beta_k)^2 = 1. \tag{6.22}
$$

Since $I([\beta])$ is continuous and the set given by (6.22) is compact, function $I([\beta])$ attains minimum, denoted by $\lambda_m^{(1)}$, at some point $[\beta^{(1)}] = (\beta_1^{(1)}, \ldots, \beta_m^{(1)})$. If this procedure is carried out for $m = 1, 2, \ldots$, we obtain a sequence of numbers $\lambda_1^{(1)}, \lambda_2^{(1)}, \ldots$. Because $\lambda_{m+1}^{(1)} \leq \lambda_m^{(1)}$ and $\mathcal{I}(y)$ is bounded from below, we can find the limit

$$
\lim_{m \to \infty} \lambda_m^{(1)} = \lambda^{(1)}.
$$

Step 2. Let

$$
y_m^{(1)}(t) = \frac{1}{\sqrt{w_\alpha}} \sum_{k=1}^{m} \beta_k^{(1)} \sin(kt)
$$

denote the linear combination (6.20) achieving the minimum $\lambda_m^{(1)}$. We shall prove that sequence $(y_m^{(1)})_{m \in \mathbb{N}}$ contains a uniformly convergent subsequence. From now on, for simplicity, we will write y_m instead of $y_m^{(1)}$. Recall that

$$
\lambda_m^{(1)} = \int_0^\pi \left[p(t) \left({}_0^C D_t^\alpha [y_m](t) \right)^2 + q(t) y_m^2(t) \right] dt
$$

is convergent, so it must be bounded, i.e., there exists a constant $M > 0$ such that

$$
\int_0^\pi \left[p(t) \left({}_0^C D_t^\alpha [y_m](t) \right)^2 + q(t) y_m^2(t) \right] dt \leq M, \quad m \in \mathbb{N}.
$$

Therefore, for all $m \in \mathbb{N}$ it hold the following:

$$\int_0^\pi p(t) \left({}_0^C D_t^\alpha [y_m](t) \right)^2 dt \leq M + \left| \int_0^\pi q(t) y_m^2(t) \, dt \right|$$

$$\leq M + \max_{t \in [0,\pi]} \left| \frac{q(t)}{w_\alpha(t)} \right| \int_0^\pi w_\alpha(t) y_m^2(t) \, dt = M + \max_{t \in [0,\pi]} \left| \frac{q(t)}{w_\alpha(t)} \right| =: M_1.$$

Moreover, since $p(t) > 0$, one has

$$\min_{t \in [0,\pi]} p(t) \int_0^\pi \left({}_0^C D_t^\alpha [y_m](t) \right)^2 dt \leq \int_0^\pi p(t) \left({}_0^C D_t^\alpha [y_m](t) \right)^2 dt \leq M_1,$$

and hence

$$\int_0^\pi \left({}_0^C D_t^\alpha [y_m](t) \right)^2 dt \leq \frac{M_1}{\min\limits_{t \in [0,\pi]} p(t)} =: M_2. \tag{6.23}$$

Using (4.67) and (6.23), condition $y_m(0) = 0$ and Schwartz's inequality, one has

$$|y_m(t)|^2 = \left| \left({}_0 I_t^\alpha \circ {}_0^C D_t^\alpha \right) [y_m](t) \right|^2 = \frac{1}{\Gamma(\alpha)^2} \left| \int_0^t (t - \tau)^{\alpha-1} {}_0^C D_\tau^\alpha y_m(\tau) \, d\tau \right|^2$$

$$\leq \frac{1}{\Gamma(\alpha)^2} \left(\int_0^\pi \left| {}_0^C D_\tau^\alpha [y_m](\tau) \right|^2 d\tau \right) \left(\int_0^t (t - \tau)^{2(\alpha-1)} \, d\tau \right)$$

$$\leq \frac{1}{\Gamma(\alpha)^2} M_2 \int_0^t (t - \tau)^{2(\alpha-1)} \, d\tau < \frac{1}{\Gamma(\alpha)} M_2 \frac{1}{2\alpha - 1} \pi^{2\alpha-1},$$

so that $(y_m)_{m \in \mathbb{N}}$ is uniformly bounded. Now, using Schwartz's inequality, Eq. (6.23) and the fact that the following inequality holds,

$$\forall x_1 \geq x_2 \geq 0, \ (x_1 - x_2)^2 \leq x_1^2 - x_2^2,$$

we have for any $0 < t_1 < t_2 \leq \pi$ that

$$|y_m(t_2) - y_m(t_1)| = \left| \left({}_0I_t^\alpha \circ {}_0^C D_t^\alpha \right) [y_m](t_2) - \left({}_0I_t^\alpha \circ {}_0^C D_t^\alpha \right) [y_m](t_1) \right|$$

$$= \frac{1}{\Gamma(\alpha)} \left| \int_0^{t_2} (t_2 - \tau)^{\alpha-1} {}_0^C D_t^\alpha [y_m](\tau) \, d\tau - \int_0^{t_1} (t_1 - \tau)^{\alpha-1} {}_0^C D_t^\alpha [y_m](\tau) \, d\tau \right|$$

$$= \frac{1}{\Gamma(\alpha)} \left| \int_{t_1}^{t_2} (t_2 - \tau)^{\alpha-1} {}_0^C D_t^\alpha [y_m](\tau) \, d\tau \right.$$

$$\left. - \int_0^{t_1} \left((t_2 - \tau)^{\alpha-1} - (t_1 - \tau)^{\alpha-1} \right) {}_0^C D_t^\alpha [y_m](\tau) \, d\tau \right|$$

$$\leq \frac{1}{\Gamma(\alpha)} \left[\left(\int_{t_1}^{t_2} (t_2 - \tau)^{2(\alpha-1)} \, d\tau \right)^{\frac{1}{2}} \left(\int_{t_1}^{t_2} \left[\left({}_0^C D_t^\alpha [y_m](\tau) \right)^2 \right] \, d\tau \right)^{\frac{1}{2}} \right.$$

$$+ \left(\int_0^{t_1} \left((t_1 - \tau)^{\alpha-1} - (t_2 - \tau)^{\alpha-1} \right)^2 \, d\tau \right)^{\frac{1}{2}}$$

$$\left. \times \left(\int_0^{t_1} \left[\left({}_0^C D_t^\alpha [y_m](\tau) \right)^2 \right] \, d\tau \right)^{\frac{1}{2}} \right]$$

$$\leq \frac{\sqrt{M_2}}{\Gamma(\alpha)} \left[\left(\int_{t_1}^{t_2} (t_2 - \tau)^{2(\alpha-1)} \, d\tau \right)^{\frac{1}{2}} \right.$$

$$\left. + \left(\int_0^{t_1} \left((t_1 - \tau)^{2(\alpha-1)} - (t_2 - \tau)^{2(\alpha-1)} \right) \, d\tau \right)^{\frac{1}{2}} \right]$$

$$= \frac{\sqrt{M_2}}{\Gamma(\alpha)\sqrt{2\alpha - 1}} \left[(t_2 - t_1)^{\alpha-\frac{1}{2}} + \left[(t_2 - t_1)^{2\alpha-1} - t_2^{2\alpha-1} + t_1^{2\alpha-1} \right]^{\frac{1}{2}} \right]$$

$$\leq \frac{2\sqrt{M_2}}{\Gamma(\alpha)\sqrt{2\alpha - 1}} (t_2 - t_1)^{\alpha-\frac{1}{2}}.$$

Therefore, by Ascoli's theorem, there exists a uniformly convergent subsequence $(y_{m_n})_{n \in \mathbb{N}}$ of sequence $(y_m)_{m \in \mathbb{N}}$. It means that we can find $y^{(1)} \in C([a, b]; \mathbb{R})$ such that

$$y^{(1)} = \lim_{n \to \infty} y_{m_n}.$$

Step 3. Observe that by the Lagrange multiplier rule at $[\beta] = [\beta^{(1)}]$ we have

$$0 = \frac{\partial}{\partial \beta_j}\left[I([\beta]) - \lambda_m^{(1)}J([\beta])\right]\big|_{[\beta]=[\beta^{(1)}]}, \quad j = 1, \ldots, m.$$

Multiplying equations by an arbitrary constant C^j and summing from 1 to m we obtain

$$0 = \sum_{j=1}^{m}C^j\frac{\partial}{\partial \beta_j}\left[I([\beta]) - \lambda_m^{(1)}J([\beta])\right]\big|_{[\beta]=[\beta^{(1)}]}. \tag{6.24}$$

Introducing

$$h_m(x) = \frac{1}{\sqrt{w_\alpha}}\sum_{j=1}^{m}C^j\sin(jt)$$

we can rewrite (6.24) in the form

$$0 = \int_0^\pi\left[p(t)_0^C D_t^\alpha[y_m](t)_0^C D_t^\alpha[h_m](t) + [q(t) - \lambda_m^{(1)}w_\alpha(t)]y_m(t)h_m(t)\right]dt. \tag{6.25}$$

Using the differentiation properties and formula $_0^C D_t^\alpha[y_m] = \frac{d}{dt}_0 I_t^{1-\alpha}[y_m]$ we write (6.25) as

$$0 = \int_0^\pi\left[-p'(t)_0 I_t^{1-\alpha}[y_m](t)_0^C D_t^\alpha[h_m](t) - p(t)_0 I_t^{1-\alpha}[y_m](t)\frac{d}{dt}_0^C D_t^\alpha[h_m](t)\right]dt$$

$$+ p(t)_0 I_t^{1-\alpha}[y_m](t)_0^C D_t^\alpha[h_m](t)\big|_{t=0}^{t=\pi}$$

$$+ \int_0^\pi[q(t) - \lambda_m^{(1)}w_\alpha(t)]y_m(t)h_m(t)\,dt := I_m. \tag{6.26}$$

By Lemma A.2 (with $w = 1/\sqrt{w_\alpha}$) and Lemma A.3 (Appendix), for function h fulfilling assumptions of Lemma 6.2, we shall obtain in the limit (at least for the convergent subsequence $(y_{m_n})_{n\in\mathbb{N}}$) the relation

$$0 = \int_0^\pi\left[-p'(t)_0 I_t^{1-\alpha}[y^{(1)}](t)\,_0^C D_t^\alpha[h](t) - p(t)_0 I_t^{1-\alpha}[y^{(1)}](t)\frac{d}{dt}_0^C D_t^\alpha[h](t)\right]dt$$

$$+ p(t)_0 I_t^{1-\alpha}[y^{(1)}](t)\,_0^C D_t^\alpha[h](t)\big|_{t=0}^{t=\pi} + \int_0^\pi[q(t) - \lambda^{(1)}w_\alpha(t)]y^{(1)}(t)h(t)\,dt := I. \tag{6.27}$$

Let us check the convergence of integrals (6.26) explicitly:

$$|I_m - I| \leq \int_0^\pi \left| -p'(t){}_0I_t^{1-\alpha}[y_m](t){}_0^C D_t^\alpha[h_m](t) + p'(t){}_0I_t^{1-\alpha}[y^{(1)}](t){}_0^C D_t^\alpha[h](t) \right| dt$$

$$+ \int_0^\pi \left| p(t){}_0I_t^{1-\alpha}[y_m](t)\frac{d}{dt}{}_0^C D_t^\alpha[h_m](t) - p(t){}_0I_t^{1-\alpha}[y^{(1)}](t)\frac{d}{dt}{}_0^C D_t^\alpha[h](t) \right| dt$$

$$+ \left| p(t){}_0I_t^{1-\alpha}[y_m](t){}_0^C D_t^\alpha[h_m](t)|_{x=0} - p(t){}_0I_t^{1-\alpha}[y^{(1)}](t){}_0^C D_t^\alpha[h](t)|_{t=0} \right|$$

$$+ \left| p(t){}_0I_t^{1-\alpha}[y_m](t){}_0^C D_t^\alpha[h_m](t)|_{t=\pi} - p(t){}_0I_t^{1-\alpha}[y^{(1)}](t){}_0^C D_t^\alpha[h](t)|_{t=\pi} \right|$$

$$+ \int_0^\pi \left| [q(t) - \lambda_m^{(1)} w_\alpha(t)]y_m(t)h_m(t) - \left[q(t) - \lambda^{(1)} w_\alpha(t)\right] y^{(1)}(t) \, h(t) \right| dt.$$

$$(6.28)$$

For the first integral we get

$$\int_0^\pi \left| -p'(t){}_0I_t^{1-\alpha}[y_m](t){}_0^C D_t^\alpha[h_m](t) + p'(t){}_0I_t^{1-\alpha}[y^{(1)}](t){}_0^C D_t^\alpha[h](t) \right| dt$$

$$\leq ||p'|| \cdot \left[||{}_0^C D_t^\alpha[h]|| \cdot ||{}_0I_t^{1-\alpha}[y_m - y^{(1)}]||_{L^1} \right.$$
$$\left. + M_3 K_{1-\alpha}\sqrt{\pi}||{}_0^C D_t^\alpha[h_m - h]||_{L^2} \right],$$

where constant $M_3 = \sup_{m\in\mathbb{N}} ||y_m||$ and $|| \cdot ||$ denotes the supremum norm in the $C([0, \pi]; \mathbb{R})$ space. Now, we estimate the second integral

$$\int_0^\pi \left| p(t){}_0I_t^{1-\alpha}[y_m](t)\frac{d}{dt}{}_0^C D_t^\alpha[h_m](t) - p(t)I_{0+}^{1-\alpha}y^{(1)}(t)\frac{d}{dt}{}_0^C D_t^\alpha[h](t) \right| dt$$

$$\leq ||p|| \cdot \left[||\frac{d}{dt}{}_0^C D_t^\alpha[h]||_{L^2} \cdot ||{}_0I_t^{1-\alpha}[y_m - y^{(1)}]||_{L^2} \right.$$
$$\left. + M_3 K_{1-\alpha} \cdot ||\frac{d}{dt}{}_0^C D_t^\alpha[h_m - h]||_{L^1} \right].$$

For the next two terms we have

$$_0I_t^{1-\alpha}[y_m](0) \xrightarrow[m\to\infty]{} {}_0I_t^{1-\alpha}[y](0), \quad {}_0I_t^{1-\alpha}[y_m](\pi) \xrightarrow[m\to\infty]{} {}_0I_t^{1-\alpha}[y](\pi) \quad (6.29)$$

resulting from the convergence of sequence $y_m \xrightarrow[m\to\infty]{C} y$. For the sequence $h_m = g_m/\sqrt{w_\alpha}$, we infer from Lemma A.3 that $h'_m \xrightarrow[m\to\infty]{C} h'$. Hence, also

$$_0^C D_t^\alpha[h_m] \xrightarrow[m\to\infty]{C} {}_0^C D_t^\alpha[h], \quad {}_0 I_t^{1-\alpha}[h'_m] \xrightarrow[m\to\infty]{C} {}_0 I_t^{1-\alpha}[h']$$

and at points $t = 0, \pi$ we obtain

$$_0^C D_t^\alpha[h_m](0) \xrightarrow[m\to\infty]{} {}_0^C D_t^\alpha[h](0), \quad {}_0^C D_t^\alpha[h_m](\pi) \xrightarrow[m\to\infty]{} {}_0^C D_t^\alpha[h](\pi). \quad (6.30)$$

The above pointwise convergence (6.29) and (6.30) imply that

$$\lim_{m\to\infty} \left| p(t){}_0 I_t^{1-\alpha}[y_m](t){}_0^C D_t^\alpha[h_m](t)|_{t=0} - p(t){}_0 I_t^{1-\alpha}[y^{(1)}](t){}_0^C D_t^\alpha[h](t)|_{t=0} \right| = 0,$$

$$\lim_{m\to\infty} \left| p(t){}_0 I_t^{1-\alpha}[y_m](t){}_0^C D_t^\alpha[h_m](t)|_{t=\pi} - p(t){}_0 I_t^{1-\alpha}[y^{(1)}](t){}_0^C D_t^\alpha[h](t)|_{t=\pi} \right| = 0.$$

Finally, for the last term in estimation (6.28) we get

$$\int_0^\pi |[q(t) - \lambda_m^{(1)} w_\alpha(t)]y_m(t)h_m(t) - [q(t) - \lambda^{(1)} w_\alpha(t)]y^{(1)}(t)\, h(t)|\, dt$$

$$\leq \int_0^\pi |q(t)(y_m(t)h_m(t) - y^{(1)}(t)\, h(t))|\, dt$$

$$+ \int_0^\pi |w_\alpha(t)(\lambda_m^{(1)} y_m(t)h_m(t) - \lambda^{(1)} y^{(1)}(t)\, h(t))|\, dt$$

$$\leq \pi \cdot \|q\| \cdot \left[M_3 \cdot \|h_m - h\| + \|h\| \cdot \|y_m - y^{(1)}\| \right]$$

$$+ \pi \cdot \|w_\alpha\| \cdot \left[\Lambda \left(M_3 \cdot \|h_m - h\| + \|h\| \cdot \|y_m - y^{(1)}\| \right) \right.$$

$$\left. + \|y^{(1)}h\| \cdot |\lambda_m^{(1)} - \lambda^{(1)}| \right],$$

where constants $M_3 = \sup_{m\in\mathbb{N}} \|y_m\|$ and $\Lambda = \sup_{m\in\mathbb{N}} |\lambda_m^{(1)}|$. We conclude that

$$0 = \lim_{m\to\infty} I_m = I$$

and (6.27) is fulfilled for function $y^{(1)}$ being the limit of subsequence (y_{m_n}) of the sequence $(y_m)_{m\in\mathbb{N}}$.

Step 4. Let us denote in relation (6.27):

$$\gamma_1(t) := (q(t) - \lambda^{(1)} w_\alpha(t))y^{(1)}(t),$$

$$\gamma_2(t) := -p'(t){}_0 I_t^{1-\alpha}[y^{(1)}](t),$$

$$\gamma_3(t) := -p(t){}_0 I_t^{1-\alpha}[y^{(1)}](t).$$

We observe that $\gamma_j \in C([0, \pi]; \mathbb{R})$, $j = 1, 2, 3$ and $_0D_t^{1-\alpha}[\gamma_3] \in L^2(0, \pi; \mathbb{R})$ because

$$_0D_t^{1-\alpha}[\gamma_3] = {}_0D_t^{1-\alpha}\left[p \cdot {}_0I_t^{1-\alpha}[y^{(1)}]\right] = {}_0I_t^{\alpha}\left[\frac{d}{dt}\left(p \cdot {}_0I_t^{1-\alpha}[y^{(1)}]\right)\right]$$

$$= {}_0I_t^{\alpha}\left[p' \cdot {}_0I_t^{1-\alpha}[y^{(1)}] + p \cdot {}_0^CD_t^{\alpha}[y^{(1)}]\right].$$

Both parts of the above function belong to the $L^2(0, \pi; \mathbb{R})$ space.

Assuming that function h in (6.27) is an arbitrary function fulfilling assumptions of Lemma 6.3 and applying Lemma 6.3 part (a), we conclude that $\gamma_3 = -p \cdot {}_0I_t^{1-\alpha}[y^{(1)}] \in C^1([0, \pi]; \mathbb{R})$. From this fact it follows that $p \cdot \frac{d}{dt}{}_0I_t^{\alpha}[y^{(1)}] \in C([0, \pi]; \mathbb{R})$ and integral (6.27) can be rewritten as

$$\int_0^\pi \left[p(t){}_0^CD_t^{\alpha}[y^{(1)}](t){}_0^CD_t^{\alpha}[h](t) + (q(t) - \lambda^{(1)}w_\alpha(t))y^{(1)}(t)h(t)\right] dt = 0.$$

Now, we apply Lemma 6.3 part (b) defining

$$\bar{\gamma}_1(t) := \gamma_1(t) = (q(t) - \lambda^{(1)}w_\alpha)y^{(1)}(t),$$

$$\bar{\gamma}_2(t) := p(t)\frac{d}{dt}{}_0I_t^{1-\alpha}[y^{(1)}](t).$$

This time $\bar{\gamma}_1, \bar{\gamma}_2 \in C([0, \pi]; \mathbb{R})$ and from Lemma 6.3 part (b) it follows that

$$_t^CD_\pi^{\alpha}\left[p(\tau){}_0^CD_\tau^{\alpha}[y^{(1)}](\tau)\right](t) + q(t)y^{(1)}(t) = \lambda^{(1)}w_\alpha(t)y^{(1)}(t).$$

By construction, this solution fulfills the Dirichlet boundary conditions

$$y^{(1)}(0) = y^{(1)}(\pi) = 0$$

and is nontrivial because

$$\mathcal{J}(y^{(1)}) = \int_0^\pi w_\alpha(t)\left(y^{(1)}(t)\right)^2 dt = 1.$$

In addition, we also have for the solution

$$_0D_t^{\alpha}[y^{(1)}] = {}_0^CD_t^{\alpha}[y^{(1)}] \in C([0, \pi]; \mathbb{R}).$$

Let us observe that from the Dirichlet boundary conditions it follows that $y^{(1)}$ also solves the FSLP (6.16) and (6.17) in $[0, \pi]$.

Step 5. Now, let us restore the superscript on $y_m^{(1)}$ and show that $\left(y_m^{(1)}\right)_{m \in \mathbb{N}}$ itself converges to $y^{(1)}$. First, let us point out that for given λ the solution of

$$_t^C D_\pi^\alpha \left[p(\tau) {_0^C} D_\tau^\alpha[y](\tau) \right](t) + q(t)y(t) = \lambda^{(1)} w_\alpha(t)y(t), \qquad (6.31)$$

subject to the boundary conditions

$$y(a) = y(b) = 0 \qquad (6.32)$$

and the normalization condition

$$\int_0^\pi w_\alpha(t)y^2(t)\,dt = 1, \qquad (6.33)$$

is unique except for a sign. Next, let us assume that $y^{(1)}$ solves the Sturm–Liouville equation (6.31) and that the corresponding eigenvalue is $\lambda = \lambda^{(1)}$. In addition, suppose that $y^{(1)}$ is nontrivial, i.e., we can find $t_0 \in [0, \pi]$ such that $y^{(1)}(t_0) \neq 0$ and choose the sign, so that $y^{(1)}(t_0) > 0$. Similarly, for all $m \in \mathbb{N}$, let $y_m^{(1)}$ solve (6.31) with corresponding eigenvalue $\lambda = \lambda_m^{(1)}$ and let us choose the signs, so that $y_m^{(1)}(t_0) \geq 0$. Now suppose that $\left(y_m^{(1)}\right)_{m \in \mathbb{N}}$ does not converge to $y^{(1)}$. It means that we can find another subsequence of $\left(y_m^{(1)}\right)_{m \in \mathbb{N}}$ such that it converges to another solution $\bar{y}^{(1)}$ of (6.31) with $\lambda = \lambda^{(1)}$. We know that for $\lambda = \lambda^{(1)}$ solution of (6.31) subject to (6.32) and (6.33) must be unique except for a sign, thence

$$\bar{y}^{(1)} = -y^{(1)}$$

and we must have $\bar{y}^{(1)}(t_0) < 0$. However, it is impossible because for all $m \in \mathbb{N}$ the value of $y_m^{(1)}$ in t_0 is greater or equal than zero. It means that we have contradiction and hence, choosing each $y_m^{(1)}$ with adequate sign, we obtain $y_m^{(1)} \to y^{(1)}$.

Step 6. In order to find eigenfunction $y^{(2)}$ and the corresponding eigenvalue $\lambda^{(2)}$, we again minimize functional (6.18) subject to (6.19) and (6.17), but now with an extra orthogonality condition

$$\int_0^\pi w_\alpha(t)y(t)y^{(1)}(t)\,dt = 0. \qquad (6.34)$$

If we approximate solution by

$$y_m(t) = \frac{1}{\sqrt{w_\alpha}} \sum_{k=1}^m \beta_k \sin(kt), \quad y_m(0) = y_m(\pi) = 0,$$

then we again receive quadratic form (6.21). However, in this case admissible solutions are points satisfying (6.22) together with

$$\frac{\pi}{2} \sum_{k=1}^{m} \beta_k \beta_k^{(1)} = 0, \tag{6.35}$$

i.e., they lie in the $(m - 1)$-dimensional sphere. As before, we find that function $I([\beta])$ has a minimum $\lambda_m^{(2)}$ and there exists $\lambda^{(2)}$ such that

$$\lambda^{(2)} = \lim_{m \to \infty} \lambda_m^{(2)},$$

because $J(y)$ is bounded from below. Moreover, it is clear that the following relation:

$$\lambda^{(1)} \leq \lambda^{(2)} \tag{6.36}$$

holds. Now, let us denote by

$$y_m^{(2)}(t) = \frac{1}{\sqrt{w_\alpha}} \sum_{k=1}^{m} \beta_k^{(2)} \sin(kt)$$

the linear combination achieving the minimum $\lambda_m^{(2)}$, where

$$\beta^{(2)} = (\beta_1^{(2)}, \ldots, \beta_m^{(2)})$$

is the point satisfying (6.22) and (6.35). By the same argument as before, we can prove that the sequence $(y_m^{(2)})_{m \in \mathbb{N}}$ converges uniformly to a limit function $y^{(2)}$, which satisfies the Sturm-Liouville equation (6.16) with $\lambda^{(2)}$, the boundary conditions (6.17), normalization condition (6.19), and the orthogonality condition (6.34). Therefore, solution $y^{(2)}$ of the FSLP corresponding to the eigenvalue $\lambda^{(2)}$ exists. Furthermore, because orthogonal functions cannot be linearly dependent, and since only one eigenfunction corresponds to each eigenvalue (except for a constant factor), we have the strict inequality

$$\lambda^{(1)} < \lambda^{(2)}$$

instead of (6.36). Finally, if we repeat the above procedure, with similar modifications, we can obtain eigenvalues $\lambda^{(3)}, \lambda^{(4)}, \ldots$ and corresponding eigenfunctions $y^{(3)}, y^{(4)}, \ldots$.

6.2.2 The First Eigenvalue

In this section we prove two theorems showing that the first eigenvalue of problem
(6.16) and (6.17) is a minimum value of certain functionals. As in the proof of
Theorem 6.5 in the sequel, for simplicity, we assume that $a = 0$ and $b = \pi$ in the
problem (6.16) and (6.17).

Theorem 6.6 *Let $y^{(1)}$ denote the eigenfunction, normalized to satisfy the isoperimetric constraint*

$$\mathcal{J}(y) = \int_0^\pi w_\alpha(t) y^2(t)\, dt = 1, \tag{6.37}$$

*associated to the first eigenvalue $\lambda^{(1)}$ of problem (6.16) and (6.17) and assume that
function ${}_tD_\pi^\alpha \left[p \cdot {}_0^C D_t^\alpha[y] \right]$ is continuous. Then, $y^{(1)}$ is a minimizer of the following
variational functional*

$$\mathcal{I}(y) = \int_0^\pi \left[p(t)({}_0^C D_t^\alpha[y](t))^2 + q(t) y^2(t) \right] dt, \tag{6.38}$$

*in the class $C([0, \pi]; \mathbb{R})$ with ${}_0^C D_t^\alpha[y] \in C([0, \pi]; \mathbb{R})$ subject to the boundary con-
ditions*

$$y(0) = y(\pi) = 0 \tag{6.39}$$

and an isoperimetric constraint (6.37). Moreover, $\mathcal{I}(y^{(1)}) = \lambda^{(1)}$.

Proof Suppose that $y \in C([0, \pi]; \mathbb{R})$ is a minimizer of $\mathcal{I}$ and ${}_0^C D_t^\alpha[y] \in C([0, \pi]; \mathbb{R})$.
Then, by Theorem 6.4, there is number λ such that y satisfies equation

$${}_tD_\pi^\alpha \left[p(\tau){}_0^C D_\tau^\alpha[y](\tau) \right](t) + q(t)y(t) = \lambda w_\alpha(t)y(t), \tag{6.40}$$

and conditions (6.37) and (6.39). Since ${}_tD_\pi^\alpha \left[p \cdot {}_0^C D_t^\alpha[y] \right]$ and ${}_t^C D_\pi^\alpha \left[p \cdot {}_0^C D_t^\alpha[y] \right]$
are continuous, it follows that $p(t) \cdot {}_0^C D_t^\alpha[y](t)\big|_{t=\pi} = 0$. Therefore, Eq. (6.40) is
equivalent to

$${}_t^C D_\pi^\alpha \left[p(\tau){}_0^C D_\tau^\alpha[y](\tau) \right](t) + q(t)y(t) = \lambda w_\alpha(t)y(t). \tag{6.41}$$

Let us multiply (6.40) by y and integrate it on the interval $[0, \pi]$. Then

$$\int_0^\pi \left(y(t) \cdot {}_tD_\pi^\alpha \left[p(\tau){}_0^C D_\tau^\alpha[y](\tau) \right](t) + q(t)y^2(t) \right) dt = \lambda \int_0^\pi w_\alpha(t)y^2(t)\, dt.$$

Applying the integration by parts formula for fractional derivatives (cf. (2.9)) and having in mind that conditions (6.39), (6.37) and

$$p(t) \cdot {}_0^C D_t^\alpha [y](t) \Big|_{t=\pi} = 0$$

hold, one has

$$\int_0^\pi \left(\left({}_0^C D_t^\alpha [y](t) \right)^2 p(t) + q(t) y^2(t) \right) dt = \lambda.$$

Hence, $\mathcal{I}(y) = \lambda$. Any solution to problem (6.37)–(6.39) that satisfies Eq. (6.41) must be nontrivial since (6.37) holds, so λ must be an eigenvalue. Moreover, according to Theorem 6.5 there is the least element in the spectrum being eigenvalue $\lambda^{(1)}$ and the corresponding eigenfunction $y^{(1)}$ normalized to meet the isoperimetric condition. Therefore $J(y^{(1)}) = \lambda^{(1)}$.

Definition 6.7 We call to functional $\mathcal{R}$ defined by

$$\mathcal{R}(y) = \frac{\mathcal{I}(y)}{J(y)},$$

where $\mathcal{I}(y)$ is given by (6.38) and $J(y)$ by (6.37), the Rayleigh quotient for the fractional Sturm–Liouville problem (6.16) and (6.17).

Theorem 6.8 *Let us assume that function $y \in C([0, \pi]; \mathbb{R})$ with ${}_0^C D_t^\alpha [y] \in C([0, \pi]; \mathbb{R})$, satisfying boundary conditions $y(0) = y(\pi) = 0$ and being nontrivial, is a minimizer of the Rayleigh quotient $\mathcal{R}$ for the Sturm–Liouville problem (6.16) and (6.17). Moreover, assume that function ${}_t D_\pi^\alpha \left[p \cdot {}_0^C D_t^\alpha [y] \right]$ is continuous. Then, the value of $\mathcal{R}$ in y is equal to the first eigenvalue $\lambda^{(1)}$, i.e., $\mathcal{R}(y) = \lambda^{(1)}$.*

Proof Suppose that function $y \in C([0, \pi]; \mathbb{R})$ with ${}_0^C D_t^\alpha [y] \in C([0, \pi]; \mathbb{R})$, satisfying $y(0) = y(\pi) = 0$ and nontrivial, is a minimizer of the Rayleigh quotient $\mathcal{R}$ and that the value of $\mathcal{R}$ in y is equal to λ, i.e.,

$$\mathcal{R}(y) = \frac{\mathcal{I}(y)}{J(y)} = \lambda.$$

Consider a one-parameter family of curves $\hat{y} = y + h\eta$, $|h| \leq \varepsilon$, where $\eta \in C([0, \pi]; \mathbb{R})$ with ${}_0^C D_t^\alpha [\eta] \in C([0, \pi]; \mathbb{R})$ is such that $\eta(0) = \eta(\pi) = 0$, $\eta \neq 0$, and define the following functions:

$$\phi : [-\varepsilon, \varepsilon] \longrightarrow \mathbb{R}$$

$$h \longmapsto J(y + h\eta) = \int_0^\pi w_\alpha(t)(y(t) + h\eta(t))^2 \, dt,$$

$$\psi : [-\varepsilon, \varepsilon] \longrightarrow \mathbb{R}$$

$$h \longmapsto \mathcal{I}(y + h\eta) = \int_0^\pi \Big[p(t)(_0^C D_t^\alpha [y + h\eta](t))^2$$

$$+ q(t)(y(t) + h\eta(t))^2 \Big]\, dt$$

and

$$\zeta : [-\varepsilon, \varepsilon] \longrightarrow \mathbb{R}$$

$$h \longmapsto \mathcal{R}(y + h\eta) = \frac{\mathcal{I}(y+h\eta)}{\mathcal{J}(y+h\eta)}.$$

Since ζ is of class C^1 on $[-\varepsilon, \varepsilon]$ and

$$\zeta(0) \leq \zeta(h), \quad |h| \leq \varepsilon,$$

we deduce that

$$\zeta'(0) = \frac{d}{dh}\mathcal{R}(y + h\eta)\Big|_{h=0} = 0.$$

Moreover, notice that

$$\zeta'(h) = \frac{1}{\phi(h)}\left(\psi'(h) - \frac{\psi(h)}{\phi(h)}\phi'(h)\right)$$

and that

$$\psi'(0) = \frac{d}{dh}\mathcal{I}(y + h\eta)\Big|_{h=0}$$

$$= 2\int_0^\pi \Big[p(t) \cdot {}_0^C D_t^\alpha [y](t) \cdot {}_0^C D_t^\alpha [\eta](t) + q(t)y(t)\eta(t) \Big]\, dt,$$

$$\phi'(0) = \frac{d}{dh}\mathcal{J}(y + h\eta)\Big|_{h=0} = 2\int_0^\pi [w_\alpha(t)y(t)\eta(t)]\, dt.$$

Therefore

$$\zeta'(0) = \frac{d}{dh}\mathcal{R}(y + h\eta)\Big|_{h=0}$$

$$= \frac{2}{\mathcal{J}(y)}\left(\int_0^\pi p(t) \cdot {}_0^C D_t^\alpha [y](t) \cdot {}_0^C D_t^\alpha [\eta](t) + q(t)y(t)\eta(t)\, dt\right.$$

$$\left. - \frac{\mathcal{I}(y)}{\mathcal{J}(y)}\int_0^\pi w_\alpha(t)y(t)\eta(t)\, dt\right) = 0.$$

Having in mind that $\frac{\mathcal{I}(y)}{\mathcal{J}(y)} = \lambda$ and $\eta(0) = \eta(\pi) = 0$, using the integration by parts formula (2.9) we obtain

$$\int_0^\pi \left({}_tD_\pi^\alpha \left[p_0^c D_\tau^\alpha[y] \right](t) + q(t)y(t) - \lambda w_\alpha(t)y(t) \right) \eta(t)\, dt = 0.$$

Now, applying the fundamental lemma of the calculus of variations, we arrive to

$$_tD_\pi^\alpha \left[p(\tau) \cdot {}_0^C D_\tau^\alpha[y](\tau) \right](t) + q(t)y(t) = \lambda w_\alpha(t)y(t). \tag{6.42}$$

Under our assumptions, $p(t) \cdot {}_0^C D_t^\alpha[y](t)\big|_{t=\pi} = 0$ and therefore Eq. (6.42) is equivalent to

$$_t^C D_\pi^\alpha \left[p(\tau) \cdot {}_0^C D_\tau^\alpha[y](\tau) \right](t) + q(t)y(t) = \lambda w_\alpha(t)y(t). \tag{6.43}$$

Since $y \neq 0$ we deduce that number λ is an eigenvalue of (6.43). On the other hand, let $\lambda^{(m)}$ be an eigenvalue and $y^{(m)}$ the corresponding eigenfunction. Then

$$_t^C D_\pi^\alpha \left[p(\tau) {}_0^C D_\tau^\alpha[y^{(m)}](\tau) \right](t) + q(t)y^{(m)}(t) = \lambda^{(m)} w_\alpha(t)y^{(m)}(t). \tag{6.44}$$

Similarly to the proof of Theorem 6.6, we can obtain

$$\frac{\int_0^\pi \left(\left({}_0^C D_t^\alpha[y^{(m)}](t) \right)^2 p(t) + q(t)(y^{(m)}(t))^2 \right) dt}{\int_0^\pi \lambda^{(m)} w_\alpha(t)(y^{(m)}(t))^2\, dt} = \lambda^{(m)},$$

for any $m \in \mathbb{N}$. That is $\mathcal{R}(y^{(m)}) = \frac{\mathcal{I}(y^{(m)})}{\mathcal{J}(y^{(m)})} = \lambda^{(m)}$. Finally, since the minimum value of $\mathcal{R}$ at y is equal to λ, i.e.,

$$\lambda \leq \mathcal{R}(y^{(m)}) = \lambda^{(m)} \quad \forall m \in \mathbb{N},$$

we have $\lambda = \lambda^{(1)}$.

6.2.3 An Illustrative Example

Let us consider the following fractional oscillator equation:

$$_tD_b^\alpha \left[{}_a^c D_\tau^\alpha[y](\tau) \right](t) - \lambda y(t) = 0, \tag{6.45}$$

where $y(a) = y(b) = 0$. One can easily check that problem of finding nontrivial solutions to Eq. (6.45) and corresponding values of parameter λ is a particular
case of problem (6.16) and (6.17) with $p(t) \equiv 1$, $q(t) \equiv 0$ and $w_\alpha(t) \equiv 1$. The
corresponding variational functional is

$$\mathcal{I}_\alpha(y) = \int_a^b p(t) \cdot ({}_a^C D_t^\alpha [y](t))^2 \, dt = ||\sqrt{p} \; {}_a^C D_t^\alpha [y]||_{L^2}^2$$

with the isoperimetric condition

$$\int_a^b y^2(t) \, dt = 1.$$

Let us fix the value of parameter p and assume that orders α_1, α_2 fulfill the condition
$\frac{1}{2} < \alpha_1 < \alpha_2 < 1$. Then, we obtain for functionals $\mathcal{I}_{\alpha_1}, \mathcal{I}_{\alpha_2}$ the following relation

$$\mathcal{I}_{\alpha_1}(y) = ||\sqrt{p} \, {}_a^C D_t^{\alpha_1}[y]||_{L^2}^2 = ||\sqrt{p} \, {}_a I_t^{1-\alpha_1} \left[\frac{d}{dt} y \right] ||_{L^2}^2$$

$$= ||\sqrt{p} \, {}_a I_t^{\alpha_2-\alpha_1} \, {}_a I_t^{1-\alpha_2} \left[\frac{d}{dt} y \right] ||_{L^2}^2$$

$$\leq K_{\alpha_2-\alpha_1}^2 \cdot ||\sqrt{p} \, {}_a^C D_t^{\alpha_2}[y]||_{L^2}^2$$

$$= K_{\alpha_2-\alpha_1}^2 \mathcal{I}_{\alpha_2}(y),$$

where we denoted

$$K_{\alpha_2-\alpha_1} := \frac{(b-a)^{\alpha_2-\alpha_1}}{\Gamma(\alpha_2 - \alpha_1 + 1)}.$$

We observe that in the above estimation two cases occur:

$$\text{if } K_{\alpha_2-\alpha_1} \leq 1, \quad \text{then } \mathcal{I}_{\alpha_1}(y) \leq \mathcal{I}_{\alpha_2}(y);$$
$$\text{if } K_{\alpha_2-\alpha_1} > 1, \quad \text{then } \mathcal{I}_{\alpha_1}(y) \leq K_{\alpha_2-\alpha_1}^2 \cdot \mathcal{I}_{\alpha_2}(y).$$

The relations between functionals for different values of fractional order lead to the
set of inequalities for eigenvalues λ_j valid for any $j \in \mathbb{N}$:

$$\text{if } K_{\alpha_2-\alpha_1} \leq 1, \quad \text{then } \lambda_j(\alpha_1) \leq \lambda_j(\alpha_2);$$
$$\text{if } K_{\alpha_2-\alpha_1} > 1, \quad \text{then } \lambda_j(\alpha_1) \leq K_{\alpha_2-\alpha_1}^2 \cdot \lambda_j(\alpha_2).$$

In particular when order $\alpha_2 = 1$ we get

$$\mathcal{I}_{\alpha_1}(y) = ||\sqrt{p_a^C} D_t^{\alpha_1}[y]||_{L^2}^2 = ||\sqrt{p_a} I_t^{1-\alpha_1}[Dy]||_{L^2}^2$$
$$\leq K_{1-\alpha_1}^2 \cdot ||\sqrt{p}Dy||_{L^2}^2 = K_{1-\alpha_1}^2 \mathcal{I}_1(y)$$

and the following relations dependent on the value of constant $K_{1-\alpha_1}$:

$$\text{if } K_{1-\alpha_1} \leq 1, \quad \text{then } \mathcal{I}_{\alpha_1}(y) \leq \mathcal{I}_1(y);$$
$$\text{if } K_{1-\alpha_1} > 1, \quad \text{then } \mathcal{I}_{\alpha_1}(y) \leq K_{1-\alpha_1}^2 \cdot \mathcal{I}_1(y).$$

Thus, comparing the eigenvalues for the fractional and the classical harmonic oscillator equation for boundary conditions $y(a) = y(b) = 0$, we conclude that the respective classical eigenvalues are higher than the ones resulting from the fractional problem for any $j \in \mathbb{N}$. Namely,

$$\text{if } K_{1-\alpha_1} \leq 1, \text{ then } \lambda_j(\alpha_1) \leq \lambda_j(1) = p\left(\frac{j\pi}{b-a}\right)^2;$$

$$\text{if } K_{1-\alpha_1} > 1, \text{ then } \lambda_j(\alpha_1) \leq K_{1-\alpha_1}^2 \cdot \lambda_j(1) = p\left(\frac{j\pi}{(b-a)^{\alpha_1}\Gamma(2-\alpha_1)}\right)^2.$$

References

Al-Mdallal QM (2009) An efficient method for solving fractional Sturm-Liouville problems. Chaos Solitons Fractals 40(1):183–189

Al-Mdallal QM (2010) On the numerical solution of fractional Sturm-Liouville problems. Int J Comput Math 87(12):2837–2845

Almeida R, Torres DFM (2011) Necessary and sufficient conditions for the fractional calculus of variations with Caputo derivatives. Commun Nonlinear Sci Numer Simul 16(3):1490–1500

Gelfand IM, Fomin SV (2000) Calculus of variations. Dover Publications Inc, New York

Klimek M, Agrawal OP (2012) On a regular fractional Sturm-Liouville problem with derivatives of order in (0, 1). In: Proceedings of the 13th international carpathian control conference, 28–31 May 2012, Vysoke Tatry (Podbanske), Slovakia, pp 284–289

Klimek M, Agrawal OP (2013a) Regular fractional Sturm-Liouville problem with generalized derivatives of order in (0,1). In: Proceedings of the IFAC joint conference: 5th SSSC, 11th WTDA, 5th WFDA, 4–6 February 2013, Grenoble, France

Klimek M, Agrawal OP (2013b) Fractional Sturm-Liouville problem. Comput Math Appl 66(5):795–812

Klimek M, Odzijewicz T, Malinowska AB (2014) Variational methods for the fractional Sturm-Liouville problem. J Math Anal Appl 416(1):402–426

Liu Y, He T, Shi H (2012) Existence of positive solutions for Sturm-Liouville BVPs of singular fractional differential equations. Politehn Univ Bucharest Sci Bull Ser A Appl Math Phys 74(1):93–108

Odzijewicz T (2013) Generalized fractional calculus of variations. PhD Thesis, University of Aveiro

Qi J, Chen S (2011) Eigenvalue problems of the model from nonlocal continuum mechanics. J Math Phys 52(7), 073516, 14 pp

van Brunt B (2004) The calculus of cariations. Springer, New York

Chapter 7
Conclusion

Abstract This book was dedicated to the generalized fractional calculus of variations. We extended standard fractional variational calculus (Almeida et al. 2015; Malinowska and Torres 2012), by considering problems with generalized fractional operators, that by choosing special kernels reduce, e.g., to fractional operators of Riemann–Liouville, Caputo, Hadamard, Riesz, or Katugampola types.

Keywords Generalized fractional operators · Generalized fractional calculus of variations · Directions for future research

First, we proved several properties of generalized fractional operators, including boundedness in the space of p-Lebesgue integrable functions, and generalized fractional integration by parts formula. Next, we applied standard methods of fractional variational calculus to find admissible functions giving minima to certain functionals. We considered cases of one and several variables. However, because in standard methods it is assumed that Euler–Lagrange equations are solvable, we presented certain results according to direct methods, where it is not the case. We proved a Tonelli-type theorem ensuring the existence of minimizers and then obtained necessary optimality condition giving candidates for solutions. The last chapter was devoted to the fractional Sturm–Liouville problem. Applying methods of fractional variational calculus, we proved that there exists an infinite increasing sequence of eigenvalues; to each eigenvalue corresponds an eigenfunction and all of them are orthogonal. Moreover, we presented two theorems concerning the first eigenvalue. Concluding, our results cover several variational problems with particular fractional operators and give a compact and transparent view for the fractional calculus of variations. We trust that our work will provide new insights to further research on the subject, where still much remains to be done. Such research can choose several directions. Here, we find important to mention the following ones. We can

- consider variational problems with higher order derivatives;
- consider Lagrangians with different operators K_{p1}, K_{p2}, ... and B_{p1}, B_{p2}, ...;
- apply direct methods to multidimensional variational problems;
- explore problems of generalized fractional calculus of variations with holonomic or nonholonomic constraints;

© The Author(s) 2015
A.B. Malinowska et al., *Advanced Methods in the Fractional Calculus of Variations*, SpringerBriefs in Applied Sciences and Technology,
DOI 10.1007/978-3-319-14756-7_7

- with the help of fractional variational calculus show fractional counterparts of isoperimetric inequality; and
- continue with variational methods in problems of generalized fractional optimal control including necessary conditions of optimality or maximum principle.

The results mentioned in this book were first published in peer reviewed international journals (Bourdin et al. 2013, 2014; Klimek et al. 2014; Odzijewicz et al. 2012a, b, c, 2013b, c, d, e; Odzijewicz and Torres 2012, 2014); chapters in books (Odzijewicz 2013b; Odzijewicz et al. 2013a); and proceedings with referee (Odzijewicz et al. 2010, 2012d, e, f). See also the PhD thesis (Odzijewicz 2013a).

References

Almeida R, Pooseh S, Torres DFM (2015) Computational methods in the fractional calculus of variations. Imperial College Press, London

Bourdin L, Odzijewicz T, Torres DFM (2013) Existence of minimizers for fractional variational problems containing Caputo derivatives. Adv Dyn Syst Appl 8(1):3–12

Bourdin L, Odzijewicz T, Torres DFM (2014) Existence of minimizers for generalized Lagrangian functionals and a necessary optimality condition—application to fractional variational problems. Differ Integral Equ 27(7–8):743–766

Klimek M, Odzijewicz T, Malinowska AB (2014) Variational methods for the fractional Sturm-Liouville problem. J Math Anal Appl 416(1):402–426

Malinowska AB, Torres DFM (2012) Introduction to the fractional calculus of variations. Imperial College Press, London

Odzijewicz T (2013a) Generalized fractional calculus of variations. PhD Thesis, University of Aveiro

Odzijewicz T (2013b) Variable order fractional isoperimetric problem of several variables. Advances in the theory and applications of non-integer order systems 257:133–139

Odzijewicz T, Malinowska AB, Torres DFM (2010) Calculus of variations with fractional and classical derivatives. In: Podlubny I, Vinagre Jara BM, Chen YQ, Feliu Batlle V, Tejado Balsera I (eds) Proceedings of FDA'10, the 4th IFAC workshop on fractional differentiation and its applications, Badajoz, Spain, 18–20 October 2010, Art. no. FDA10-076, 5 pp

Odzijewicz T, Malinowska AB, Torres DFM (2012a) Generalized fractional calculus with applications to the calculus of variations. Comput Math Appl 64(10):3351–3366

Odzijewicz T, Malinowska AB, Torres DFM (2012b) Fractional variational calculus with classical and combined Caputo derivatives. Nonlinear Anal 75(3):1507–1515

Odzijewicz T, Malinowska AB, Torres DFM (2012c) Fractional calculus of variations in terms of a generalized fractional integral with applications to physics. Abstr Appl Anal 2012(871912), 24 pp

Odzijewicz T, Malinowska AB, Torres DFM (2012d) Green's theorem for generalized fractional derivatives. In: Chen W, Sun HG, Baleanu D (eds) Proceedings of FDA'2012, the 5th symposium on fractional differentiation and its applications, 14–17 May 2012, Hohai University, Nanjing, China. Paper #084

Odzijewicz T, Malinowska AB, Torres DFM (2012e) A Generalized fractional calculus of variations with applications. In: Proceedings of the 20th international symposium on mathematical theory of networks and systems (MTNS), 9–13 July 2012, University of Melbourne, Australia, Paper 159

Odzijewicz T, Malinowska AB, Torres DFM (2012f) Variable order fractional variational calculus for double integrals. In: Proceedings of the IEEE conference on decision and control 6426489:6873–6878

Odzijewicz T, Malinowska AB, Torres DFM (2013a) Fractional variational calculus of variable order. Advances in harmonic analysis and operator theory, Operator theory: advances and applications, vol 229. Birkhäuser, Basel, pp 291–301

Odzijewicz T, Malinowska AB, Torres DFM (2013b) Green's theorem for generalized fractional derivative. Fract Calc Appl Anal 16(1):64–75

Odzijewicz T, Malinowska AB, Torres DFM (2013c) A generalized fractional calculus of variations. Control Cybern 42(2):443–458

Odzijewicz T, Malinowska AB, Torres DFM (2013d) Fractional calculus of variations of several independent variables. Eur Phys J Spec Top 222(8):1813–1826

Odzijewicz T, Malinowska AB, Torres DFM (2013e) Noether's theorem for fractional variational problems of variable order. Cent Eur J Phys 11(6):691–701

Odzijewicz T, Torres DFM (2012) Calculus of variations with classical and fractional derivatives. Math Balkanica 26(1–2):191–202

Odzijewicz T, Torres DFM (2014) The generalized fractional calculus of variations. Southeast Asian Bull Math 38(1):93–117

Appendix A
Two Convergence Lemmas

In this appendix we prove two lemmas, concerning certain convergence properties of fractional and classical derivatives.

Lemmas A.2 and A.3 are important in the proof of Theorem 6.5. Let us begin with the following definition of Hölder continuous functions.

Definition A.1 Function g is Hölder continuous in the interval $[a, b]$ with coefficient $0 < \beta \leq 1$ if

$$\sup_{x,y\in[a,b],\ x\neq y} \frac{|g(x) - g(y)|}{|x - y|^\beta} < \infty. \tag{A.1}$$

We denote this class of Hölder continuous functions as $C_H^\beta([a, b]; \mathbb{R})$.

Lemma A.2 Let $\alpha \in (0, 1)$, functions $w, g \in C^1([0, \pi]; \mathbb{R}) \cap C_H^1([-\pi, \pi]; \mathbb{R})$ be odd functions in $[-\pi, \pi]$ such that $w'', g'' \in L^2(0, \pi; \mathbb{R})$. If we denote as g_m the mth sum of the Fourier series of function g, then the following convergences are valid in $[0, \pi]$:

$$\substack{C \\ 0} D_t^\alpha[g_m] \xrightarrow[m\to\infty]{L^2} \substack{C \\ 0} D_t^\alpha[g], \tag{A.2}$$

$$\frac{d}{dt} \substack{C \\ 0} D_t^\alpha[g_m] \xrightarrow[m\to\infty]{L^1} \frac{d}{dt} \substack{C \\ 0} D_t^\alpha[g], \tag{A.3}$$

$$\substack{C \\ 0} D_t^\alpha[wg_m] \xrightarrow[m\to\infty]{L^2} \substack{C \\ 0} D_t^\alpha[wg], \tag{A.4}$$

$$\frac{d}{dt} \substack{C \\ 0} D_t^\alpha[wg_m] \xrightarrow[m\to\infty]{L^1} \frac{d}{dt} \substack{C \\ 0} D_t^\alpha[wg]. \tag{A.5}$$

© The Author(s) 2015
A.B. Malinowska et al., *Advanced Methods in the Fractional Calculus of Variations*, SpringerBriefs in Applied Sciences and Technology, DOI 10.1007/978-3-319-14756-7

Proof We can apply Property 2.11 and estimate the $\left\| {}_0^C D_t^\alpha [g_m] - {}_0^C D_t^\alpha [g] \right\|_{L^2}$ norm in $[0, \pi]$ as follows:

$$\left\| {}_0^C D_t^\alpha [g_m] - {}_0^C D_t^\alpha [g] \right\|_{L^2} \leq \left\| {}_0 I_t^{1-\alpha} [g'_m - g'] \right\|_{L^2} \leq K_{1-\alpha} \cdot \|g'_m - g'\|_{L^2}.$$

For even functions from the $C^1([0, \pi]; \mathbb{R}) \cap C_H^1([-\pi, \pi]; \mathbb{R})$ space, g'_m is the mth sum of the Fourier series of the derivative g'. Hence, $g'_m \xrightarrow[m\to\infty]{L^2} g'$ in interval $[0, \pi]$ and from the above inequalities it follows that (A.2) is valid on $[0, \pi]$:

$$ {}_0^C D_t^\alpha [g_m] \xrightarrow[m\to\infty]{L^2} {}_0^C D_t^\alpha [g]. $$

Let us observe that for $t > 0$:

$$ \frac{d}{dt} {}_0^C D_t^\alpha [g_m](t) = {}_0 D_t^\alpha [g'_m](t) = {}_0^C D_t^\alpha [g'_m](t) + \frac{g'_m(0) \cdot t^{-\alpha}}{\Gamma(1-\alpha)}, $$

$$ \frac{d}{dt} {}_0^C D_t^\alpha [g](t) = {}_0 D_t^\alpha [g'](t) = {}_0^C D_t^\alpha [g'](t) + \frac{g'(0) \cdot t^{-\alpha}}{\Gamma(1-\alpha)}. $$

Therefore, we can estimate the distance between $\frac{d}{dt} {}_0^C D_t^\alpha [g_m]$ and $\frac{d}{dt} {}_0^C D_t^\alpha [g]$ in interval $[0, \pi]$ using (2.10) for $\beta = 1 - \alpha$ and $p = 1$:

$$\left\| \frac{d}{dt} {}_0^C D_t^\alpha [g_m] - \frac{d}{dt} {}_0^C D_t^\alpha [g] \right\|_{L^1}$$

$$\leq \left\| {}_0^C D_t^\alpha [g'_m - g'] \right\|_{L^1} + \left\| (g'_m(0) - g'(0)) \cdot \frac{t^{-\alpha}}{\Gamma(1-\alpha)} \right\|_{L^1}$$

$$= \left\| {}_0 I_t^{1-\alpha} [g''_m - g''] \right\|_{L^1} + |g'_m(0) - g'(0)| \cdot \left\| \frac{t^{-\alpha}}{\Gamma(1-\alpha)} \right\|_{L^1}$$

$$\leq K_{1-\alpha} \cdot \|g''_m - g''\|_{L^1} + |g'_m(0) - g'(0)| \cdot \frac{\pi^{1-\alpha}}{\Gamma(2-\alpha)}$$

$$\leq K_{1-\alpha} \cdot \sqrt{\pi} \cdot \|g''_m - g''\|_{L^2} + |g'_m(0) - g'(0)| \cdot \frac{\pi^{1-\alpha}}{\Gamma(2-\alpha)}.$$

By assumptions, we have in $[-\pi, \pi]$ (thence also in $[0, \pi]$) that

$$ g''_m \xrightarrow[m\to\infty]{L^2} g'', \quad g'_m(0) \xrightarrow[m\to\infty]{} g'(0). $$

Hence, we conclude that (A.3) is valid. The convergence given in (A.4) follows from (A.2), namely

$$\left\| {}^C_0 D^\alpha_t[wg_m] - {}^C_0 D^\alpha_t[wg] \right\|_{L^2}$$

$$= \left\| {}_0 I^{1-\alpha}_t \left[(wg_m)' - (wg)' \right] \right\|_{L^2}$$

$$\leq \left\| {}_0 I^{1-\alpha}_t [w(g'_m - g')] \right\|_{L^2} + \left\| {}_0 I^{1-\alpha}_t [w'(g_m - g)] \right\|_{L^2}$$

$$\leq K_{1-\alpha} \cdot \|w(g'_m - g')\|_{L^2} + K_{1-\alpha} \cdot \|w'(g_m - g)\|_{L^2}$$

$$\leq K_{1-\alpha} \left(\|w\| \cdot \|g'_m - g'\|_{L^2} + \|w'\| \cdot \|g_m - g\|_{L^2} \right),$$

where $\| \cdot \|$ denotes the supremum norm in the $C([0, \pi]; \mathbb{R})$ space. From assumptions of our lemma, it follows that $g'_m \xrightarrow[m\to\infty]{L^2} g'$ and $g_m \xrightarrow[m\to\infty]{L^2} g$ in $[0, \pi]$. Thus, convergence (A.4) is valid. To prove convergence (A.5) we start by observing that for $t > 0$ we have:

$$\frac{d}{dt} {}^C_0 D^\alpha_t[wg_m](t) = {}_0 D^\alpha_t \left[(wg_m)' \right](t) = {}^C_0 D^\alpha_t \left[(wg_m)' \right](t) + \frac{(wg_m)'(0) \cdot t^{-\alpha}}{\Gamma(1-\alpha)},$$

$$\frac{d}{dt} {}^C_0 D^\alpha_t[wg](t) = {}_0 D^\alpha_t \left[(wg)' \right](t) = {}^C_0 D^\alpha_t \left[(wg)' \right](t) + \frac{(wg)'(0) \cdot t^{-\alpha}}{\Gamma(1-\alpha)}.$$

For the L^1-distance between $\frac{d}{dt} {}^C_0 D^\alpha_t[wg_m]$ and $\frac{d}{dt} {}^C_0 D^\alpha_t[wg]$ in the interval $[0, \pi]$,

$$\left\| \frac{d}{dt} {}^C_0 D^\alpha_t[wg_m] - \frac{d}{dt} {}^C_0 D^\alpha_t[wg] \right\|_{L^1}$$

$$\leq \| {}^C_0 D^\alpha_t \left[(wg_m)' - (wg)' \right] \|_{L^1} + \left\| \left[(wg_m)'(0) - (wg)'(0) \right] \cdot \frac{t^{-\alpha}}{\Gamma(1-\alpha)} \right\|_{L^1}$$

$$\leq \| {}_0 I^{1-\alpha}_t \left[(wg_m)'' - (wg)'' \right] \|_{L^1} + \left| (wg_m)'(0) - (wg)'(0) \right| \cdot \left\| \frac{t^{-\alpha}}{\Gamma(1-\alpha)} \right\|_{L^1}$$

$$\leq K_{1-\alpha} \cdot \| (wg_m)'' - (wg)'' \|_{L^1} + \left\| (wg_m)'(0) - (wg)'(0) \right\| \cdot \frac{\pi^{1-\alpha}}{\Gamma(2-\alpha)}$$

$$\leq K_{1-\alpha} \cdot \sqrt{\pi} \cdot \| (wg_m)'' - (wg)'' \|_{L^2} + \left\| (wg_m)'(0) - (wg)'(0) \right\| \cdot \frac{\pi^{1-\alpha}}{\Gamma(2-\alpha)}.$$

$$\tag{A.6}$$

Because

$$(wg_m)'' - (wg)'' = w(g''_m - g'') + 2(w)' \cdot (g'_m - g') + (w)'' \cdot (g_m - g)$$

we have

$$\| (wg_m)'' - (wg)'' \|_{L^2}$$
$$\leq \|w\| \cdot \|g''_m - g''\|_{L^2} + 2 \cdot \| (w)' \| \cdot \|g'_m - g'\|_{L^2} + \|w''\|_{L^2} \cdot \|g_m - g\|_{L^2}.$$

From the assumptions of the lemma, it follows that for $j = 0, 1, 2$

$$\lim_{m \longrightarrow \infty} ||g_m^{(j)} - g^{(j)}||_{L^2} = 0.$$

Hence,

$$\lim_{m \longrightarrow \infty} || (wg_m)'' - (wg)'' ||_{L^2} = 0.$$

In addition,

$$\lim_{m \longrightarrow \infty} |(wg_m)'(0) - (wg)'(0)|$$

$$= \lim_{m \longrightarrow \infty} |(w)'(0)(g_m(0) - g(0)) + w(0)(g_m'(0) - g'(0))|$$

$$\leq \lim_{m \longrightarrow \infty} |(w)'(0)(g_m(0) - g(0))| + \lim_{m \longrightarrow \infty} |w(0)(g_m'(0) - g'(0))| = 0.$$

Taking into account estimation (A.6) and the above inequalities, we conclude that (A.5) is valid.

Lemma A.3 Let $\alpha \in \left(\frac{1}{2}, 1\right)$, $0 < \beta \leq \alpha - \frac{1}{2}$, function w be positive, even function in $[-\pi, \pi]$ and $w' \in C_H^\beta([-\pi, \pi]; \mathbb{R})$. Function h' is the derivative of h defined by assumptions of Lemma 6.2 and formula (6.7), function g is defined as

$$g(t) := h(t)w(t).$$

If we denote as g_m the mth sum of the Fourier series of function g, then the following convergences are valid in interval $[0, \pi]$:

$$g_m' \xrightarrow[m \to \infty]{C} g', \tag{A.7}$$

$$g_m'(0) \xrightarrow[m \to \infty]{} g'(0), \tag{A.8}$$

$$g_m'(\pi) \xrightarrow[m \to \infty]{} g'(\pi). \tag{A.9}$$

Proof Definition (6.7) in interval $[0, \pi]$ implies for derivative h' that

$$h'(t) = {}_0I_t^\alpha[\gamma](t) + At^\alpha + Bt^{1+\alpha}, \tag{A.10}$$

where $\gamma \in C[0, \pi]$ and constants $A, B \in \mathbb{R}$ are specified by conditions (6.3), (6.4) in the proof of Lemma 6.1. Let us observe that $t^{1+\alpha} \in C^1([0, \pi]; \mathbb{R})$, function t^α is Hölder continuous in $[0, \pi]$ with coefficient $\beta \leq \alpha$, thus it can be extended to an odd/even, Hölder continuous function in interval $[-\pi, \pi]$. In addition $_0I_t^\alpha[\gamma](t)$ is Hölder continuous function in $[0, \pi]$ with coefficient $\beta \leq \alpha - \frac{1}{2}$ because:

$$\frac{\left|_0I_t^\alpha[\gamma](t) - _0I_t^\alpha[\gamma](s)\right|}{|t - s|^\beta} \leq \frac{2 \cdot \|\gamma\|_{L^2}}{\Gamma(\alpha)\sqrt{2\alpha - 1}} \cdot |t - s|^{\alpha - \frac{1}{2} - \beta}$$

$$\leq \frac{2 \cdot \|\gamma\|_{L^2}}{\Gamma(\alpha)\sqrt{2\alpha - 1}} \cdot \pi^{\alpha - \frac{1}{2} - \beta} < \infty$$

and can be extended to an odd/even, Hölder continuous function in interval $[-\pi, \pi]$. Observe that for Hölder continuous functions in $[-\pi, \pi]$ we have the absolute convergence of their Fourier series. For function g' we obtain in $[0, \pi]$:

$$g'(t) = h'(t)w(t) + h(t)w'(t).$$

Both terms on the right-hand side are, by assumption, functions from the $C_H^\beta([0, \pi]; \mathbb{R})$ space and can be extended to odd/even functions in $C_H^\beta([-\pi, \pi]; \mathbb{R})$ space. Hence, their Fourier series are absolutely convergent in $[-\pi, \pi]$. Concluding, we have for function g' the convergence in interval $[-\pi, \pi]$:

$$g_m' \xrightarrow[m \to \infty]{C} g'.$$

Thus the sequence g_m' of partial sums is also absolutely convergent in the interval $[0, \pi]$. Formulas (A.8) and (A.9) are a straightforward consequence of this fact.

Index

Symbols

A_P, 15

A_{P_i}, 19

B_P, 15

B_{P_i}, 19

$C_H^\beta([a,b]; \mathbb{R})$, 127

$E_\alpha(z)$, 40

K_{P_i}, 19

K_P, 14

${}_a^C D_t^\alpha$, 9

${}_{a_i}^C D_{t_i}^{\alpha_i}$, 17

${}_t^C D_b^\alpha$, 9

${}_{t_i}^C D_{b_i}^{\alpha_i}$, 17

${}_a D_t^\alpha$, 9

${}_{a_i} D_{t_i}^{\alpha_i}$, 17

${}_t D_b^\alpha$, 9

${}_{t_i} D_{b_i}^{\alpha_i}$, 17

${}_a J_t^\alpha$, 8

${}_t J_b^\alpha$, 8

${}_a I_t^\alpha$, 8

${}_{a_i} I_{t_i}^{\alpha_i}$, 17

${}_t I_b^\alpha$, 8

${}_{t_i} I_{b_i}^{\alpha_i}$, 17

$\mathscr{P}_M$, 86

$\mathbf{D}_i$, 77

$\mathbf{I}_i$, 77

$\mathcal{I}$, 36

$\mathcal{N}_\delta(\bar{y})$, 31

∇_{D^α}, 69

∇_D, 70

$\nabla_{I^{1-\alpha}}$, 69

∇_I, 70

∇_{T_P}, 66

$\partial_i F$, 24

${}_{a_i}^C D_{t_i}^{\alpha_i(\cdot,\cdot)}$, 18

${}_a^C D_t^{\alpha(\cdot,\cdot)}$, 13

${}_{t_i}^C D_{b_i}^{\alpha_i(\cdot,\cdot)}$, 18

${}_t^C D_b^{\alpha(\cdot,\cdot)}$, 13

${}_a D_x^\alpha$, 2

${}_a I_t^\alpha$, 2

${}_a^C D_x^\alpha$, 3

${}_{a_i} D_{t_i}^{\alpha_i(\cdot,\cdot)}$, 18

${}_{a_i} I_{t_i}^{\alpha_i(\cdot,\cdot)}$, 18

${}_a D_t^{\alpha(\cdot,\cdot)}$, 13

${}_a I_t^{\alpha(\cdot,\cdot)}$, 12

${}_{t_i} D_{b_i}^{\alpha_i(\cdot,\cdot)}$, 18

${}_{t_i} I_{b_i}^{\alpha_i(\cdot,\cdot)}$, 18

${}_t D_b^{\alpha(\cdot,\cdot)}$, 13

${}_t I_b^{\alpha(\cdot,\cdot)}$, 13

B

Boundedness
    of generalized fractional integral K_P, 33
    of Riemann–Liouville fractional integral, 11, 33
    of variable order fractional integral, 34

C

Caldirola–Kanai Lagrangian, 62

Caputo
    derivatives of variable order, 13
    fractional derivatives, 9

© The Author(s) 2015

A.B. Malinowska et al., *Advanced Methods in the Fractional Calculus of Variations*, SpringerBriefs in Applied Sciences and Technology, DOI 10.1007/978-3-319-14756-7

partial
   derivatives of variable order, 18
partial fractional derivatives, 17

D
Dirichlet's principle, 71

E
Euler–Lagrange equation
   for problems with Caputo fractional
      derivatives, 42, 50, 61
   for problems with generalized fractional
      operators, 37, 47, 54, 60, 90
   for problems with Riemann–Liouville
      fractional integrals and Caputo frac-
      tional derivatives, 40
   for problems with variable order frac-
      tional integrals and derivatives, 42, 53
   of several variables
      for problems with generalized frac-
      tional operators, 67, 73
      for problems with Riemann–
      Liouville fractional integrals and
      Caputo fractional derivatives, 69, 75
      for problems with variable order frac-
      tional operators, 70, 75

F
Fractional operator of order (α, β), 28

G
Generalized fractional
   derivatives
      of Caputo type, 15
      of Riemann–Liouville type, 15
   integrals of Riemann–Liouville type, 14
Generalized fractional Dirichlet's principle,
   72
Generalized partial fractional
   derivative
      of Caputo type, 19
      of Riemann–Liouville type, 19
   integral, 19

H
Hadamard fractional integrals, 8

I
Integration by parts formula

for fractional derivatives, 11
for fractional integrals of Riemann–
   Liouville type, 36
for operator K_P, 35
for variable order fractional integrals, 36
of several variables
   for fractional derivatives, 65
   for generalized fractional derivatives,
   65
   for generalized fractional integrals,
   63
   for Riemann–Liouville fractional
   integrals, 64
   for variable order fractional derivatives, 66
   for variable order fractional integrals,
   65
Invariant Lagrangian
   for with generalized fractional operators,
   56
   of several variables with generalized
   fractional operators, 77

N
Natural boundary conditions
   for generalized fractional operators, 45
   for problems with Caputo fractional
      derivatives, 45
   for problems with variable order frac-
      tional integrals and derivatives, 45
Noether's Theorem
   for fractional Caputo derivatives, 58
   for generalized fractional operators, 56
   for Riemann–Liouville fractional integrals
      and Caputo fractional derivatives, 59
   for variable order fractional integrals and
      derivatives, 58
   of several variables, 78
      for generalized fractional operators,
      77
      for problems with Riemann–
      Liouville fractional integrals and
      Caputo fractional derivatives, 78

R
Rayleigh quotient, 117
Relation between
   Riemann–Liouville and Caputo frac-
      tional derivatives, 10
Riemann–Liouville
   derivatives of variable order, 13
   fractional derivatives, 9

fractional integrals, 8
integrals of variable order, 12
partial
    derivatives of variable order, 18
    integrals of variable order, 18
partial fractional derivatives, 17
partial fractional integrals, 17

S
Set $\mathscr{P}_M$, 86
Sturm–Liouville problem of fractional order,
    105

Sufficient condition for regular Lagrangian,
    86
Symmetric fractional derivatives, 25

T
The soap bubble problem, 24
Time reversible coherence, 29
Tonelli-type theorem, 84

W
Wave equation, 69
Weak dissipative parameter, 62